TABLE OF CONTENTS
TABLE DES MATIERES

SESSION 1 - SÉANCE 1

Chairman - Président : B. LOPEZ PEREZ (Spain)

SESSION 2 - SÉANCE 2

Chairman - Président : M. PEEHS (Federal Republic of Germany)

SESSION 3 - SÉANCE 3

Chairman - Président : E.O. MAXWELL (United Kingdom)

SESSION 4 - SÉANCE 4

Chairman - Président : A.B. JOHNSON (United States)

DRY STORAGE
OF SPENT FUEL ELEMENTS

STOCKAGE A SEC DES
ELEMENTS COMBUSTIBLES
IRRADIES

Proceedings of an NEA Specialist Workshop
Compte rendu d'une réunion de spécialistes de l'AEN

MADRID
11-13 May 1982

organised by the
OECD NUCLEAR ENERGY AGENCY
in collaboration with the
JUNTA DE ENERGIA NUCLEAR

organisée par
l'AGENCE DE L'OCDE POUR L'ÉNERGIE NUCLÉAIRE
en collaboration avec la
JUNTA DE ENERGIA NUCLEAR

NUCLEAR ENERGY AGENCY
ORGANISATION FOR ECONOMIC CO-OPERATION AND DEVELOPMENT

AGENCE POUR L'ÉNERGIE NUCLÉAIRE
ORGANISATION DE COOPÉRATION ET DE DÉVELOPPEMENT ÉCONOMIQUES

The Organisation for Economic Co-operation and Development (OECD) was s⌐
under a Convention signed in Paris on 14th December, 1960, which provides that the OEC
shall promote policies designed:
— to achieve the highest sustainable economic growth and employment and a risi⌐
standard of living in Member countries, while maintaining financial stability, and
thus to contribute to the development of the world economy;
— to contribute to sound economic expansion in Member as well as non-member
countries in the process of economic development;
— to contribute to the expansion of world trade on a multilateral, non-discriminatory
basis in accordance with international obligations.
The Members of OECD are Australia, Austria, Belgium, Canada, Denmark,
Finland, France, the Federal Republic of Germany, Greece, Iceland, Ireland, Italy,
Japan, Luxembourg, the Netherlands, New Zealand, Norway, Portugal, Spain, Sweden,
Switzerland, Turkey, the United Kingdom and the United States.

The OECD Nuclear Energy Agency (NEA) was established on 20th April 1972, replac-ing OECD's European Nuclear Energy Agency (ENEA) on the adhesion of Japan as a full Member.

NEA now groups all the European Member countries of OECD and Australia, Canada, Japan, and the United States. The Commission of the European Communities takes part in the work of the Agency.

The primary objectives of NEA are to promote co-operation between its Member governments on the safety and regulatory aspects of nuclear development, and on assessing the future role of nuclear energy as a contributor to economic progress.

This is achieved by:
— *encouraging harmonisation of governments' regulatory policies and practices in the nuclear field, with particular reference to the safety of nuclear installations, protection of man against ionising radiation and preservation of the environment, radioactive waste management, and nuclear third party liability and insurance;*
— *keeping under review the technical and economic characteristics of nuclear power growth and of the nuclear fuel cycle, and assessing demand and supply for the different phases of the nuclear fuel cycle and the potential future contribution of nuclear power to overall energy demand;*
— *developing exchanges of scientific and technical information on nuclear energy, particularly through participation in common services;*
— *setting up international research and development programmes and undertakings jointly organised and operated by OECD countries.*

In these and related tasks, NEA works in close collaboration with the International Atomic Energy Agency in Vienna, with which it has concluded a Co-operation Agreement, as well as with other international organisations in the nuclear field.

L'Organisation de Coopération et de Développement Économiques (OCDE), qui a été instituée par une Convention signée le 14 décembre 1960, à Paris, a pour objectif de promouvoir des politiques visant :

— à réaliser la plus forte expansion possible de l'économie et de l'emploi et une progression du niveau de vie dans les pays Membres, tout en maintenant la stabilité financière, et contribuer ainsi au développement de l'économie mondiale ;
— à contribuer à une saine expansion économique dans les pays Membres, ainsi que non membres, en voie de développement économique ;
— à contribuer à l'expansion du commerce mondial sur une base multilatérale et non discriminatoire, conformément aux obligations internationales.

Les Membres de l'OCDE sont : la République Fédérale d'Allemagne, l'Australie, l'Autriche, la Belgique, le Canada, le Danemark, l'Espagne, les États-Unis, la Finlande, la France, la Grèce, l'Irlande, l'Islande, l'Italie, le Japon, le Luxembourg, la Norvège, la Nouvelle-Zélande, les Pays-Bas, le Portugal, le Royaume-Uni, la Suède, la Suisse et la Turquie.

L'Agence de l'OCDE pour l'Énergie Nucléaire (AEN) a été créée le 20 avril 1972, en remplacement de l'Agence Européenne pour l'Énergie Nucléaire de l'OCDE (ENEA) lors de l'adhésion du Japon à titre de Membre de plein exercice.

L'AEN groupe désormais tous les pays Membres européens de l'OCDE ainsi que l'Australie, le Canada, les États-Unis et le Japon. La Commission des Communautés Européennes participe à ses travaux.

L'AEN a pour principaux objectifs de promouvoir, entre les gouvernements qui en sont Membres, la coopération dans le domaine de la sécurité et de la réglementation nucléaires, ainsi que l'évaluation de la contribution de l'énergie nucléaire au progrès économique.

Pour atteindre ces objectifs, l'AEN :

— *encourage l'harmonisation des politiques et pratiques réglementaires dans le domaine nucléaire, en ce qui concerne notamment la sûreté des installations nucléaires, la protection de l'homme contre les radiations ionisantes et la préservation de l'environnement, la gestion des déchets radioactifs, ainsi que la responsabilité civile et les assurances en matière nucléaire ;*
— *examine régulièrement les aspects économiques et techniques de la croissance de l'énergie nucléaire et du cycle du combustible nucléaire, et évalue la demande et les capacités disponibles pour les différentes phases du cycle du combustible nucléaire, ainsi que le rôle que l'énergie nucléaire jouera dans l'avenir pour satisfaire la demande énergétique totale ;*
— *développe les échanges d'informations scientifiques et techniques concernant l'énergie nucléaire, notamment par l'intermédiaire de services communs ;*
— *met sur pied des programmes internationaux de recherche et développement, ainsi que des activités organisées et gérées en commun par les pays de l'OCDE.*

Pour ces activités, ainsi que pour d'autres travaux connexes, l'AEN collabore étroitement avec l'Agence Internationale de l'Énergie Atomique de Vienne, avec laquelle elle a conclu un Accord de coopération, ainsi qu'avec d'autres organisations internationales opérant dans le domaine nucléaire.

Following removal from a reactor on completion of their operational life, the spent fuel elements are stored under water for several months to allow radioactive decay of short-lived radioisotopes and initial cooling. Subsequenly, the spent fuel may be reprocessed as soon as technically feasible or stored for a further period pending a decision on whether it is to be reprocessed or otherwise disposed of. The duration of this storage and the ultimate disposition of the spent fuel will depend on national policies regarding reprocessing and recovery of usable fuel. But whatever the ultimate solution adopted, some form of interim storage will be necessary, whether at or away from the reactor site.

In June 1978, recognizing the need to bring together the experience gained in OECD Member countries in the area of spent fuel management, the NEA organized, in collaboration with the Junta de Energia Nuclear, a Technical Seminar which was held in Madrid and provided a comprehensive review of spent fuel storage technology as it then existed. At that time the prevailing technique considered was wet storage.

It is widely accepted that most fuel cycle strategies, even given a fairly rapid expansion of reprocessing capability will involve a considerable and increasing transient backlog of stored spent fuel, and since the early 1970s there has been growing interest and activity in various Member countries in the development of dry storage techniques, which do not involve very complex technologies and appear to offer several technical and economic advantages over wet storage. Within the past few years several dry storage concepts have been developed to the stage of commercial production.

The NEA therefore readily accepted the invitation of the Vice-President of the Spanish Junta de Energia Nuclear to organize, in collaboration with the Junta, a further meeting on the storage of spent fuel elements, which would recognize and emphasize the emerging new developments in dry storage techniques. The Specialist Workshop was arranged accordingly and discussed the present state of the art of dry storage, future trends, and research and development work.

Après avoir été retirés d'un réacteur à l'issue de leur durée de vie utile, les éléments combustibles irradiés sont stockés sous eau pendant plusieurs mois, afin de permettre la décroissance radioactive des radioisotopes de courte période et le refroidissement initial. Par la suite, le combustible irradié peut être retraité dès que cela s'avère possible sur le plan technique ou être stocké pendant une nouvelle période dans l'attente d'une décision sur la question de savoir s'il doit être retraité ou évacué d'une autre façon. La durée de ce stockage et l'évacuation définitive du combustible irradié dépendront des politiques nationales à l'égard du retraitement et de la récupération du combustible utilisable. Cependant, quelle que soit la solution adoptée en fin de compte, il faut prévoir quelque forme de stockage provisoire, soit sur le site du réacteur, soit à distance de ce site.

Vu la nécessité de mettre en commun l'expérience acquise dans les pays Membres de l'OCDE en matière de gestion du combustible irradié, l'AEN a organisé, en juin 1978 à Madrid, en collaboration avec la Junta de Energia Nuclear, un séminaire technique qui a permis de dresser un état complet de la technologie du stockage du combustible irradié. A l'époque, la principale technique envisagée était le stockage sous eau.

On s'accorde à reconnaître que, même dans l'hypothèse d'une expansion relativement rapide de la capacité théorique de retraitement, la plupart des stratégies liées au cycle du combustible nucléaire impliqueront une accumulation temporaire considérable et croissante de combustibles irradiés stockés, aussi la mise au point de techniques de stockage à sec qui ne font pas intervenir des technologies très complexes et paraissent offrir plusieurs avantages techniques et économiques par rapport au stockage sous eau suscite-t-elle, depuis le début des années 70, de plus en plus d'intérêt et d'activités dans divers pays Membres. C'est ainsi que, ces dernières années, plusieurs modes de stockage à sec ont atteint le stade de l'exploitation commerciale.

En conséquence, l'AEN a volontiers accepté l'invitation qui lui était faite par le Vice-Président de la Junta de Energia Nuclear (Espagne) d'organiser, en collaboration avec cette dernière, une nouvelle réunion sur le stockage des éléments combustibles irradiés destinée à prendre en compte et à mettre en lumière les progrès qui se font jour dans les techniques de stockage à sec. La réunion de spécialistes a été organisée dans cet esprit afin d'examiner l'état actuel des connaissances relatives au stockage à sec, les tendances futures, ainsi que les travaux de recherche et de développement.

PANEL DISCUSSION - TABLE RONDE

Chairman - Président : H.R. KONVICKA (Austria)

WELCOME ADDRESS

Professor Dr. Baldomero LOPEZ PEREZ
Director of the Back End of the Nuclear Fuel Cycle
Junta de Energia Nuclear (Spain)

Ladies and Gentlemen,

I should like to express, both on behalf of the Junta de Energia Nuclear and myself, a warm welcome to you to this Specialist Workshop on Techniques for the Dry Storage of Spent Fuel Elements, organized in Madrid, by the OECD Nuclear Energy Agency in collaboration with the Junta de Energia Nuclear (JEN).

As you know, on April 1977, President Carter made public his nuclear policy, in which an indefinite moratorium in the reprocessing of spent nuclear fuel was imposed. Due to this decision, storage of spent fuel, that up to that moment had been considered as a transitory step in the back end of the nuclear fuel cycle, was automatically given increased emphasis.

The existing know-how in that field, that is, the storage of spent fuel under water, was a process with limited duration and did not present serious problems. Then, however, the idea of post-poning or even eliminating the reprocessing of the spent fuel gave rise to the concept of long term storage and doubts began to emerge as to whether this technology would be able to support longer periods of time without deterioration of the chemical and mechanical stability of the spent fuel elements.

This preoccupation was perceived in all the occidental countries with nuclear installations and encouraged the OECD Nuclear Energy Agency to organize a Seminar which was held in Madrid in June 1978, with the collaboration of the Junta de Energia Nuclear. At that meeting, the main subject was under-water storage systems and the exchange of points of view proved fruitful. Everybody agreed to maintain these contacts periodically.

At the same time, the International Nuclear Fuel Cycle Evaluation (INFCE) was also meeting, in which, as you know, Group VI that I have the honour to co-chair with Dr. Antonio Carrea, was reviewing the basic concepts of the transportation and storage of spent fuel. The Group's studies included existing know-how, techniques under development, safety, environmental impact, economic considerations, institutional analysis, etc. It is worth pointing out that from the information gathered, it was considered that towards the year 2000 the production of spent fuel would be of the order of 300 000 tons of heavy metal.

The INFCE did not make a choice between under-water or dry technology since at that time, dry storage was not yet considered as a proven technology. However, this does not mean that dry storage technology was not in use, as Canada has stored its CANDU type fuel elements in the dry state from the beginning.

The INFCE Final Plenary Conference was held in February 1980, and in the post-INFCE period, several international activities have continued to maintain different studies in the spent fuel management field. At the same time, individual countries have developed new policies and ideas, looking for their own best solution of the various problems, having regard to their particular technical, economic, financial, safety and political situation.

Within this international framework, an Expert meeting on Alternatives for Storage of Spent Fuel was held in Las Vegas, organized by the U.S. Department of Energy in November 1980. At this meeting, several countries reported on their research concerning dry storage techniques, which are already at an advanced stage of realization either as pilot plant facilities or as systems in the course of being licensed. For the first time, it was concluded that dry storage could begin to be considered as a proven technology.

Also within the framework of international collaboration, and in recognition of the importance attached to the storage and transportation steps in the back end of the nuclear fuel cycle, the IAEA in 1979 started an International Program of Spent Fuel Management. A Group of Experts was created whose prime objectives were the study of a potential international collaboration in the field of spent fuel management, as well as to provide assistance and an assessment to the Agency. As you know, this Group of Experts was afterwards divided into two sub-groups A and B. The first one carried out work on technical and economic considerations; the second one was concerned with the institutional and legal problems.

Both sub-groups, Group A by mid 1981 and Group B at the beginning of 1982, published their conclusions. In summary, the main conclusions were that up to 1990 there would be no major problems regarding the management of spent fuel, since the requirements for storage could be amply covered. Several technologies were identified; some, like pools, already known and proved but which nevertheless should continue to be developed, especially to increase storage density, and others that seemed to be very promising like dry techniques but whose actual state of development was less known. The economic study made allowed the costs of the various concepts to be compared and the effects of the different parameters to be determined. With regard to institutional considerations, the Group recommended that work and cooperation at international level should continue, encouraging the different concepts of regional, multi-national or international facilities, which through bilateral or international agreements, could contribute to the development of solutions for spent fuel management.

In this review of the effort devoted at international level to spent fuel management, I should not forget the study started in 1977 on the long term behaviour of spent fuel with high burn-up. This programme, whose original name was COFAST (Corrosion of Spent Fuel Assemblies during Long-Term Storage) was later calles BEFAST, changing the word "corrosion" to "behaviour", since the latter included more diversified aspects than just corrosion. The programme is still going on, under the auspices of the OECD.

Finally, I would like to indicate that in the last inter-national ENS/ANS Conference, held in Brussels, two weeks ago, on "New Directions in Nuclear Energy with Emphasis on Fuel Cycles", one of the sessions was devoted to Spent Fuel Storage, and one of the panels to "What to do with Spent Fuel", including spent fuel management without reprocessing.

Ladies and Gentlemen, I would like to finish this introduction to our meeting by saying that it has been organized with the same spirit of collaboration to which I have referred during my talk, and its purpose is to present and discuss the advances made in the subject since the last meeting in Las Vegas. We believe that our consideration only of dry techniques fits in well with the new trends in the field of spent fuel storage. If we consider that in under-water storage practical experience has been successful, we may also agree that in dry storage there remain further developments to be undertaken. I remember something a French soldier said, when he received a difficult order : "Ce qui est possible est fait; ce qui est impossible sera fait" (What is possible has been done; what is impossible will be done). I think this should be our approach.

I repeat our welcome to you. I am confident that this meeting will lead to good results, and I am sure that you will enjoy your stay in Madrid.

Thank you.

WELCOME ADDRESS

W.T. POTTER
Nuclear Development Division
OECD Nuclear Energy Agency

I would first like to thank Professor Lopez Perez on behalf of us all, for his welcome and to thank the Junta for the admirable spirit of international collaboration which has enabled this meeting to be arranged in Madrid. We particularly appreciate the excellent facilities made available to us.

On behalf of Mr. Howard Shapar, Director General of the OECD Nuclear Energy Agency I am happy to welcome you all and wish to say at once how pleased we are that this meeting has attracted so much interest.

It is widely accepted that most fuel cycle strategies, even given a fairly rapid expansion of reprocessing capability, will involve a considerable and increasing transient backlog of stored spent fuel, and there has been growing interest and activity in various of our Member countries in the development of dry storage techniques, based on their economic potential. NEA, therefore, responded readily when at the beginning of last year Senor Francisco Pascual Martinez, Vice-President and Director General of the Junta, extended his kind invitation to organize, in collaboration with the Junta, a further meeting on the storage of spent fuel elements, which would recognize and emphasize the emerging developments in dry storage techniques. This specialist workshop was arranged accordingly.

When we began to consider the organisational arrangements for the meeting we accepted that although the topic for discussion was important, it was also very specialized, and we expected to have about 30 or 40 participants at the most. However, in the event, the response far exceeded our expectations and we have rather more than 90 participants registered for the Workshop, coming from 13 countries together with the IAEA and NEA.

This gratifying response fully justifies our belief that it would be very timely to investigate the present state of the art in dry storage techniques in some detail. We have an interesting programme before us and I am sure that the various papers will lead to fruitful discussions both inside and outside the workshop and that the meeting will be of great interest and benefit to all participants as well as to those who will read the proceedings in due course.

Ladies and Gentlemen, the Nuclear Energy Agency is dedicated to international cooperation, and this workshop is a practical demonstration of what we set out to achieve in many such meetings arranged as part of our programme. Here is the opportunity for specialists from different countries to discuss together their problems and ideas on how best to solve them. I hope that you will all make the most of his opportunity and that fruitful relationships established now will encourage further international contacts and cooperation.

I wish all participants a successful meeting and look forward to the pleasure of seeing you at the reception we shall be having this evening.

SESSION 1

Chairman - Président

B. LOPEZ PEREZ

(Spain)

SEANCE 1

LONG-TERM DRY STORAGE DEMONSTRATION

WITH SPENT LWR FUEL

J. Fleisch, K. Einfeld and A. Lührmann
Deutsche Gesellschaft für Wiederaufarbeitung
von Kernbrennstoffen mbH
P.O. Box 1407, D-3000 Hannover 1,
Federal Republic of Germany

ABSTRACT

Experience with dry storage of spent fuel in several countries has indicated the potential of this technology to provide a safe and in particular economic alternative interim storage concept. Among the several dry storage techniques operating worldwide, the concept of above-ground cask storage was selected in Germany.

The program now underway is integrated with a series of single fuel pin tests as well as whole fuel bundle tests. The first operational data with a fully instrumented mobile storage cask of the CASTOR type are discussed. The major objective of the cask program is to expand the already worldwide available data base on fuel performance on a larger statistical scale. The experimental data base presented confirm the dry storage cask concept and the spent fuel storage conditions.

1. INTRODUCTION AND BASIC PHILOSOPHY

The decision to reprocess spent nuclear fuel from commercial light
water reactors has been delayed in the Federal Republic of Germany
for political reasons, except the Federal and state's Goverments
resolution to demonstrate technical-scale operation of a German reprocessing plant with a 350 t U/a capacity. Assuming besides the
amount of spent fuel contracted to the French company COGEMA for
reprocessing, the operation of a German reprocessing plant with a
capacity of 350 t U/a in mid 1990, and less restricted licensing
policy   for installation of compact storage racks in nuclear power
plant pools, there is still even for the 'minimum-prognostication'
of installed nuclear power capacity an amount of about 5.000 t U of
spent fuel left in the year 2.000. From that it can be concluded,
that a safe and in particular economic interim storage strategy is
essential before finally closing the nuclear fuel cycle.

Under these assumptions an alternative interim storage technique
based on the cask concept was developed. Already in October 1979
the first licensing procedure for an 'away-from-reactor (AFR)'
storage facility with a capacity of 1.500 t U was initiated by DWK.

The design criteria for a mobile storage cask are the normal type B
(U) licensing criteria as established by the IAEA and for storage
purposes the requirement of resistance to aircraft impacts and in
order to fulfil long term aspects, the realization of a double
barrier containment.

To achieve these criteria DWK and STEAG jointly developed a cast nodular iron cask, which has already been tested extensively and demonstrated that it fulfils the requirements for transport as well as
storage. The engineering was executed by the Gesellschaft für Nuklear-Service (GNS),an affiliate of STEAG, and VEBA-Kraftwerke Ruhr in
cooperation with Siempelkamp foundry. The result is a series of casks
of the 'CASTOR' type, holding up to 9 PWR or 16 BWR fuel assemblies
with at least 4.8 t U [1]. The first B (U) license for a cask out of
the whole CASTOR-family was already issued on May 1980.

2. CURRENT STATE OF DRY STORAGE APPLICABLE TO FUEL FROM COMMERCIAL LIGHT WATER REACTORS (LWR)

2.1 Operating Experience with Dry Storage Systems

In the frame of a 'Dry Storage Technology Development Program' sponsored by the U.S. Department of Energy (DOE) with the objective to
expand the technology data base required for the principal concepts:
drywells, surface casks (concrete) and air cooled vaults, an experience base has been provided for LWR fuel [2].

Different dry storage systems have been successfully demonstrated
with the following test results, important for dry storage concepts
in general:

- Passive heat transfer has been verified for near-surface (drywell)
  and above-ground (cask) storage configurations without causing any
  damage to the fuel.

- Due to the practical shielding design only very low personnel radiation exposures have been observed during dry fuel assembly handling.

- Dry fuel assembly handling has not caused any radioactive contamination problems.

In order to expand the dry storage experience base, DWK in cooperation with one of the German utilities, Preussenelektra (PE), started
a demonstration program with a mobile storage cask at the Würgassen
nuclear power plant early 1982. Table I lists the operation history
with dry storage systems for LWR fuel and a few characteristic data.
It does not contain the status and extensive operating experience
with vault and cask type systems for dry storage of MAGNOX and CANDU
type fuel.

2.2 Long-Term Spent Fuel Performance

Spent fuel behaviour does not affect the CASTOR cask concept, because even under the hypothetical assumption of a 100 % pin rupture
the cask and its barrier system constitutes a safe containment.
However, with regard to the fuel unloading procedure at the back end
of the fuel cycle that question might be of importance; e.g. for
operation of a fuel receiving station in a reprocessing plant.

Theoretical investigations on long-term spent fuel behaviour under
dry storage conditions have shown /6/ that cladding rupture is caused
first and foremost by the creep deformation of the fuel rods due
to their internal pressure. The temperature dependence of that failure mechanism is evident. To verify these calculations, single fuel
rod and whole fuel bundle tests are performed in the frame of a
joint R&D program with NUKEM and Kraftwerk Union (KWU). The bundle
experiment will be presented to that workshop by KWU.

Table II shows characteristics data from elevated temperature testing
of whole spent fuel rods in the range 400 to 571 °C done by BATTELLE,
Columbus, Labs. /3/ and NUKEM together with EURATOM, Ispra /4/. From
a first preliminary interpretation of the pre- and post-test fuel
rod mechanical data of a 10 month testing with 6 months at 400 °C
isothermal, it can be concluded that under cask storage conditions
with an insertion temperature of 400 °C the experimental creep strain
is much less than 0.1 %. The corresponding Kr-85 release data of
$< 8.5 \cdot 10^{-7}$ Ci/m^3 indicate, that no pin leaker occured during the
storage period.

Even in the temperature region 482 to 571 °C no cladding breaches
occured even though the tests operated much longer in time than
their lifetime was predicted theoretically. From that it can be
concluded that 400 °C is a conservative value for long-term cladding
integrity.

3. HANDLING AND DEMONSTRATION WITH AN INSTRUMENTED TRANSPORT AND STORAGE CASK OF THE CASTOR TYPE

3.1 Program Overview

A test program overview as well as the corresponding program steps
are roughly shown in Figure 1. The general aims of a CASTOR test
with a fully instrumented cask and fuel bundles are:

- to verify experimentally thermal properties and to calibrate
  thermal computer models;

- to verify and confirm the general radiation shielding characteristics;

- to obtain experience in cask handling, cask maintenance and service techniques as well as personnel training and

Table I Operation History with Dry Storage Systems for Spent LWR Fuel

Storage Configuration	Location		Fuel Type, No of Assemblies	Heat Generation per Unit	Operation History						Reference
					1978	1979	1980	1981	1982	1983	
Transport and Storage Cask (CASTOR-Type)	Würgassen Power Plant		BWR 16	19 kw							
Concrete Cask (Sealed Storage Cask)	Nevada Test Site	E-MAD[1]	PWR 1	0.9 kw							[2]
Drywell (Soil)		E-MAD	PWR 2	0.9 kw							[2]
Air Cooled Vault		E-MAD	PWR (11)	2.0 kw							[2]
Drywell (Granite)		CLIMAX	PWR 11	2.4 kw							[2]

[1]) Engine - Maintenance, Assembly and Disassembly Facility

Table II Characteristic Data of Fuel Rod Testing under Elevated Dry Storage Conditions

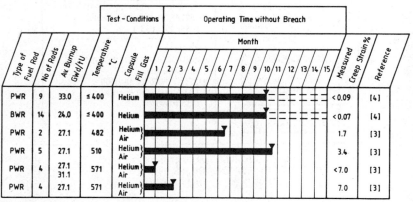

Type of Fuel Rod	No of Rods	Av Burnup Gwd/tU	Temperature °C	Capsule Fill Gas	Operating Time without Breach (Month)	Measured Creep Strain %	Reference
PWR	9	33.0	≤ 400	Helium		< 0.09	[4]
BWR	14	24.0	≤ 400	Helium		< 0.07	[4]
PWR	2	27.1	482	Helium) Air }		1.7	[3]
PWR	5	27.1	510	Helium) Air }		3.4	[3]
PWR	4	27.1 31.1	571	Helium) Air }		< 7.0	[3]
PWR	4	27.1	571	Helium) Air }		7.0	[3]

▼ Post-Test Examination
‐‐ continued

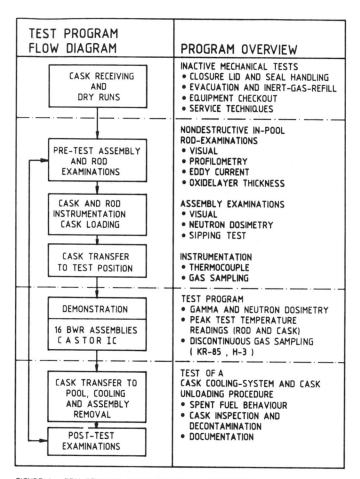

TEST PROGRAM FLOW DIAGRAM	PROGRAM OVERVIEW
CASK RECEIVING AND DRY RUNS	INACTIVE MECHANICAL TESTS • CLOSURE LID AND SEAL HANDLING • EVACUATION AND INERT-GAS-REFILL • EQUIPMENT CHECKOUT • SERVICE TECHNIQUES
PRE-TEST ASSEMBLY AND ROD EXAMINATIONS	NONDESTRUCTIVE IN-POOL ROD-EXAMINATIONS • VISUAL • PROFILOMETRY • EDDY CURRENT • OXIDELAYER THICKNESS
CASK AND ROD INSTRUMENTATION CASK LOADING	ASSEMBLY EXAMINATIONS • VISUAL • NEUTRON DOSIMETRY • SIPPING TEST
CASK TRANSFER TO TEST POSITION	INSTRUMENTATION • THERMOCOUPLE • GAS SAMPLING
DEMONSTRATION 16 BWR ASSEMBLIES C A S T O R IC	TEST PROGRAM • GAMMA AND NEUTRON DOSIMETRY • PEAK TEST TEMPERATURE READINGS (ROD AND CASK) • DISCONTINUOUS GAS SAMPLING (KR-85 , H-3)
CASK TRANSFER TO POOL, COOLING AND ASSEMBLY REMOVAL POST-TEST EXAMINATIONS	TEST OF A CASK COOLING-SYSTEM AND CASK UNLOADING PROCEDURE • SPENT FUEL BEHAVIOUR • CASK INSPECTION AND DECONTAMINATION • DOCUMENTATION

FIGURE 1   DRY STORAGE DEMONSTRATION IN TRANSPORT CASKS :
          TEST PROGRAM FLOW DIAGRAM AND OVERVIEW

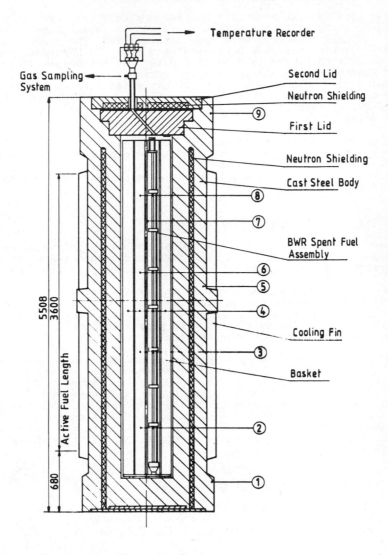

Temperature Recorder

Gas Sampling System

Second Lid

Neutron Shielding

⑨

First Lid

Neutron Shielding

Cast Steel Body

⑧

⑦

BWR Spent Fuel Assembly

⑥

⑤

④

Cooling Fin

③

Basket

②

①

5508

3600

Active Fuel Length

680

①－⑨   Thermocouple Locations

Figure 2   Cask and Fuel Assembly Instrumentation: Axial Thermocouple Locations

- to expand the technical data base on fuel performance on a larger statistical scale.

3.2 Pre-Test Fuel Assembly and Rod Examinations

Before loading the cask with hot fuel, a cold test program was performed. Parallel to these cold tests, 16 BWR spent fuel assemblies and selected fuel rods were pre-characterized. The assemblies which are subject to the cask test originate from the Würgassen nuclear power station, which is a boiling water reactor operated by PE. All assemblies resided in the core during 4 cycles of operation, burning up to about 27.8 GWd/t U. They were selected by the following criteria: no pin leakers, high average burnup, representative load during reactor operation and bundles with 7x7 as well as 8x8 rod array.

The pre-test examinations include besides an overall assembly leak sipping test, a series of fuel rod tests e.g. visual documentation, dimensional, eddy current and oxide coating measurements. Some fabricational and pre-test examination statistics are listed in Table III.

3.3 Cask and Fuel Bundle Instrumentation

Figure 2 shows a schematic drawing of the cask design and the axial thermocouple locations. The cask first and second lid contains a thermocouple tube for insertion on the fuel assembly instrumentation.

The cask, therefore, was modified to permit installation for temperature instrumentation and gas sampling. Under an angle of 45° a boring was drilled into the first lid, and a stainless steel pipe inserted and welded to the lid. The pipe is oriented in a way to minimize gamma and neutron dose rate.

Because there are already temperature data available from full scale heater tests, the number of radial thermocouples could be limited and optimized corresponding to the scheme in Figure 3. Thermocouples are located at fuel pins, basket and inner cask wall. The fuel instrumentation is restricted to two cask sections and four fuel assemblies, which provides sufficient data on the overall temperature profile. The thermocouples are fixed to the fuel rods using an O-ring type clamp, which is in direct contact to the cladding surface.

3.4 Test Results and Experiences

The cask loading procedure with an instrumented fuel bundle is shown in Figure 4. Before lifting the cask from the bottom of the storage pool with the reactor crane, the thermocouples were fitted to the first lid.

The next steps include evacuating the inner cask cavity, backfilling with Helium and fastening the second lid. Then the thermocouples were soldered to an instrumentation plate. The metal seal checked step by step Helium leak and the second lid applied and the covers definitely fastened. The temperature recorder and the gas sampling system were connected after the cask has been transferred to its final test position.

Collection of data during cask handling and the long-term test period includes primarily a continous reading of all thermocouple

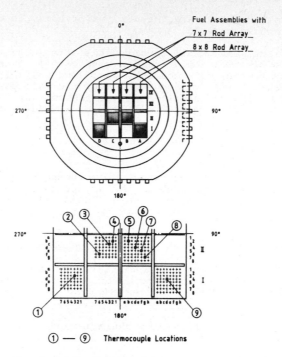

①——⑨ Thermocouple Locations

Figure 3 Fuel Assembly Instrumentation :
Radial Thermocouple Locations

Table III Fabrication and Pre-Test Fuel Rod and
Assembly Examination Statistics

Fabrication Statistics	Fuel Assembly			
	Type		BWR	BWR
	Length	mm	4470	4470
	Rod Array		7×7	8×8
	Fuel Rod			
	Number		49	63
	Length	mm	4140	4140
	Outer Diameter	mm	14.3	12.5
	Material		Zry-2	Zry-2
	Fuel Pellet			
	Material		UO_2	UO_2
	Enrichment	%	2.6	2.6
Operating History	Irradiation Time	d	2022	2022
	Average Burnup	MWd/tU	27,800	27,800
	Cycles of Operation	No.	4	4
	Cooling Time	a	1	1
	Thermal Power	kW	1.2	1.2
Pre-Test Examinations	Assembly Leak Sipping		negative	
	Fuel Rod Characteristics			
	Outer Diameter	mm	14.28 (av.)	12.46 (av.)
	Oxide Thickness μm		normal	
	Eddy Current Testing		no cladding discontinuities	

data and gas sampling (e.g. Kr-85) periodically to detect, if any
pin leakers have occured.

A detailed analysis of all experimental data (e.g. radial and axial
temperature distribution, γ- and n-dose rates) is in progress. The
preliminary thermal data could be described sufficiently taking into
account heat-transfer via convection of the fill gas Helium. The
calculated γ- and n-dose rates /5/ on the cask and lid surfaces are
well in the expected range or in case of the lid surfaces less by a
factor of about 10.

4. DISCUSSION

The first operational experience and corresponding test results confirm the made assumptions about the dry storage cask concept and the
cask loading and handling procedure. In addition the technology data
base for operating an interim storage plant could be expanded.

- Wet loading in a reactor storage pool and specific handling of a
  dry storage cask has been successfully demonstrated.

- The passive heat transfer capabilities of the CASTOR cask have
  been verfied. The maximum local fuel rod temperatures do not exceed the predicted temperatures even for fuel with about one year
  decay time.

- The total radiation shielding characteristics are verified in
  practice.

5. ACKNOWLEDGMENT

The program is supported by DWK, which is in cooperation with
Preussenelektra leading the project. KWU and GNS have carried
out significant parts of the program under contract to DWK.

The authors would like to thank K. Ramcke, H. Cmok from PE and
D. Methling from GNS for their efforts in preparing the program.
The Würgassen staff and KWU, H. Hünner and G. Kaspar, for cask
handling, instrumentation and fuel examination is also greatly
appreciated.

The fuel pin tests of chapter 2 were performed by D. Stahl and
G. Porsch from NUKEM GmbH.

6. REFERENCES

[1] Dierkes, P., Janberg, K., Baatz H. and Weinhold, G.:
     "Transport Casks Help Solve Spent Fuel Interim Storage
     Problems", Nucl. Eng. Int., October 1980

[2] Wright, J.B.: "Spent Fuel Dry Storage, A Look at the
     Past, Present and Future", Fuel Cycle Conf., Atomic
     Industrial Forum Inc., Los Angeles, (1981)

[3] Einzinger, R.E., Atkin, S.D., Stellrecht, D.E. and
     Pasupathi, V.: "High Temperature Post-Irradiation
     Materials Performance of Spent Pressurized Water
     Reactor Fuel Rods under Dry Storage Conditions",
     Nucl. Techn., to be published

[4] Stahl, D., Fleisch, J. and Porsch, D.:
     "Hot Cell Experiments for Dry Storage of Spent LWR
     Fuel", Int. ENS/ANS Conf., Brüssel, 1982

[5] Denk, W.: "Shield Design for a New Generation of
     Shipping Casks", Proc. 6 th Int. Symp. Packaging
     and Transportation of Radioactive Materials, pp.
     1172-1175, Berlin, 1980

[6] Peehs, M., Kaspar, G., Jung, W. and Schlemmer, F.:
     "Long- Term Storage Behaviour of Spent LWR Fuel",
     Proc. 6 th Int. Symp. Packaging and Transportation
     of Radioactive Materials, pp. 939-952, Berlin, 1980

Figure 4 Cask wet Loading with
Instrumented Spent Fuel Bundle

DISCUSSION

A.B. Johnson, United States

Does your programme include assessment of crud behaviour in the Würgassen cask test ?

J. Fleisch, Federal Republic of Germany

At the end of the demonstration period the cask will be unloaded. For that purpose a cask cooling system will be applied. The equipment allows samples to be taken out of the flooded cask periodically during the cask cooling operation. Information on crud behaviour could be gained from a specific analysis. In addition, experience from wet unloading of dry shipping casks collected in the Karlsruhe reprocessing plant gives information on that question.

A.B. Johnson, United States

When the CASTOR casks are used for commercial storage, will any monitoring be required (for example, temperature, ^{85}Kr, O_2) ?

J. Fleisch, Federal Republic of Germany

There is no need for monitoring, e.g. cask temperatures, radioactivity release or oxygen, because the cask and its double barrier system constitute a safe containment even under severe hypothetical accidents. The only parameter to be monitored continuously during commercial storage is the interspace pressure (pressure between first and second lid, which is above atmospheric) to control the lid system integrity carefully.

E.O. Maxwell, United Kingdom

I find the demonstration test with live fuel very laudable particularly with the complex handling requirements necessary for loading the flask. I would like to ask the speaker if he had taken into consideration carrying out these tests with electrically heated elements to simulate the heat loading and thus simplifying the tests ?

J. Fleisch, Federal Republic of Germany

A series of cask tests with electrical heaters have been performed on model casks. The test results are the basis for the arrangement of thermocouple locations selected for the hot cask demonstration and the predicted cask and fuel temperature profiles. As I stated in my paper, one of the objectives of a cask demonstration program is to verify these inactive test results, e.g. the passive decay heat capabilities.

J. Haddon, United Kingdom

Was the Gorleben site licence application made by DWK ?

J. Fleisch, Federal Republic of Germany

Yes, the licence application for the 15 000 MTU dry storage facility at the Gorleben site was made by DWK.

J. Haddon, United Kingdom

Does it specifically include CASTOR flask storage ?

J. Fleisch, Federal Republic of Germany

The Gorleben facility is based on different types of CASTOR casks, designed for PWR as well as BWR spent fuel. Other cask types that fulfil the design criteria mentioned in my presentation could be included, e.g. casks from the TN series. The casks must be licensed for transport as well as storage.

DEVELOPMENT OF A DRY TRANSPORT AND STORAGE CASK

FOR SPENT LWR FUEL ASSEMBLIES IN SPAIN

by

C. Melches   (ENUSA)
A. Uriarte   (JEN)
J.A. Espallardo (ENSA) et al.

ABSTRACT

One of the advantages of the cask storage concept is its flexibility which makes it specially attractive in the case of the Spanish circumstances. For these reasons the Empresa Nacional del Uranio, S.A. (ENUSA),Junta de Energía Nuclear (JEN) and Equipos Nucleares, S.A. (ENSA) initiated in 1981 a joint program for the development of a prototype cask for the dry transport and storage of spent fuel assemblies.

This program includes as main steps the analysis of the conceptual design, the detailed design and experimental tests, the fabrication of a prototype and its licencing and safety testing.

The mentioned program, which started in the early 1981, is scheduled to be completed at the end of 1984.

RESUMÉ

Un des principaux avantages du concept de stockage à sec des combustibles irradiés est sa flexibilité, qui le rend particulièrement adapté au cas espagnol ; c'est pour cette raison que les Entreprises Empresa Nacional del Uranio, S.A. (ENUSA), Junta de Energia Nuclear (JEN) et Equipos Nucleares, S.A. (ENSA) ont commencé, en 1981, un programme conjoint de développement d'un prototype de château pour le transport et le stockage d'éléments combustibles irradiés (LWR).

Ce programme comprend l'étude des spécifications de conception, un projet détaillé et des essais expérimentaux, la fabrication d'un prototype, son autorisation et les essais de sûreté.

Le programme, qui a commencé au début de 1981, devrait être arrêté à la fin de 1984.

INTRODUCTION

The Spanish Plan Energético Nacional (National Energy Plan) (1) of 1979 established the following strategy for the management of spent nuclear fuel:

a) As short term measure, the increase of the spent fuel storage capacity at the reactors.

b) On medium term, the construction of centralized away - from - reactor storage facilities, in order to ensure the normal operation of Spanish nuclear power plants.

c) On medium and long term, to take the necessary steps to achieve the objective of a national independence in this key area of the nuclear fuel cycle.

Wet spent fuel storage in pools appeared, until recently, the most adequate solution and almost the only available. More recently, due to the considerable development work accomplished on dry storage techniques, on the one hand, and to the consecutive delays of the Spanish nuclear program makes spent fuel storage in casks specially attractive. The cummulative need of AFR storage capacity, in the period of 1990 - 2000, will be approximately 1500 tU. In addition, this solution is inherently very flexible, specially in the sense of enabling to adjust the storage capacity closely to the real needs.

For these reasons and according to the above mentioned policy of the Plan Energético Nacional, the Empresa Nacional del Uranio, S.A. (ENUSA), Junta de Energía Nuclear (JEN) and Equipos Nucleares, S.A. (ENSA) initiated in 1981 a joint project for the development of a prototype cask for dry storage spent LWR fuel assemblies.

This paper reviews some of the different aspects of this program which includes the analysis of the conceptual design, the detailed design and model testing, the fabrication of a prototype and its safety testing.

The mentioned program, which started in the early 1981 is scheduled to be completed at the end of 1984.

2. FUNCTIONAL SPECIFICATIONS
(A. Martínez - Esparza and E. Ramírez, ENUSA)

The main objective is the of design a container suitable for dry storage of spent nuclear LWR fuel. The container shall comply with the requirements of the Fissile Class II, Type B (U) as defined in the IAEA regulations (2), as well as the applicable regulations and standards for storage of spent nuclear fuel.

The total cask weight shall not exceed 120 t, fully loaded. The cask will not exceed 500 cm in length and 200 cm in diameter. The cask shall be designed for wet loading and unloading.

Fuel cladding temperature will not exceed 250º C, which is

considered rather conservative value.

The maximum external dose rates under storage conditions will not exceed 40 mrem/hr at the external surface of the cask.

The first prototype cask shall be designed to contain 17 PWR fuel assemblies of the standard 17 x 17 PWR type design (fig.1).

The initial design enrichment shall be 3.7% U-235. The design burn - up shall be 40,000 MWd/tU; the minimum cooling time shall be 5 years.

3. RADIATION SOURCES
(A. Esteban, JEN)

The radiation sources have been calculated using the ORIGEN computer code (3).

3.1. NEUTRON EMISSION SOURCES

For burn - up higher than 20.000 Mwd/tU, Cm 244 is the major contributing element to the neutron emission rate.

Neutron sources are due to spontaneous fission of Plutonium, Curium and Californium. Moreover, neutron sources from (α, n) reactions are considered. It should be borne in mind that Pu 238, Am 241 and Cm 244 are alpha emitters and (α, n) reactions of high energy alpha particles with light elements constitute an important source of neutrons. However, the main neutron source is spontaneous fission.

The following values have been calculated:

(α, n) reactions: 1.31×10^7 neutrons / sec / assembly
Spontaneous fission: 3.45×10^8 neutrons / sec / assembly
Total: 3.58×10^8 neutrons / sec / assembly.

3.2. GAMMA SOURCES

The out - put of ORIGEN gives photon sources from fission products, structural materials (including cladding) and actinides. The photons are divided into a number of energy intervals (from 0,3 MeV to 5,25 MeV).

Results have been as follows:

Photons/sec (fission products): 0.55×10^{16}/assem.
Photons/sec (actinides): 1.32×10^{13}/assem.
Photons/sec (structural and cladding) 1.55×10^{13}/assem.

Photons from fission products are the main gamma source.

3.3. DECAY HEAT

There are three sources of heat generation: fission pro - ducts, decay of actinides and cladding/structural materials.

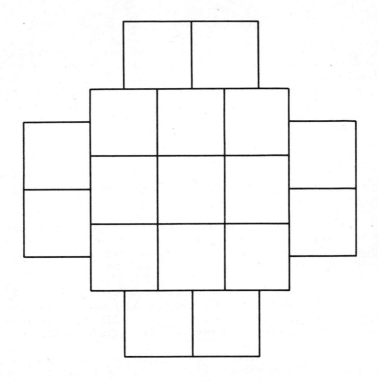

FIG. I.- DISTRIBUTION OF 17 PWR SPENT FUELS IN THE
BASKET.

C. Melches et al.

The following results have been obtained:

Fission products:	1,120 W/assembly
Decay actinides:	209 W/assembly
Structural / cladding:	5 W/assembly
Total:	1,334 W/assembly

4. CRITICALITY ANALYSIS
(J. Peña and J.M. Perlado, JEN)

The criticality analysis of the proposed cask will be performed considering loading conditions, with active PWR fuel enriched 3.7 % U-235, in the reactor pool with pure water. Boron carbide neutron poison will be used to obtain a K-effective $\leq$ 0.95. Criticality control will be achieved by aluminium or stainless steel tubes filled with compacted boron carbide, arranged in holes between adjacent fuel assemblies running the length of the basket. The alternative use of boral channels is also being considered.

An additional criticality calculation will be performed for dry storage conditions with 3.7% U-235 enrichment, assuming that the poison has disappeared.

The nuclear criticality analysis is being made using the discrete ordinates TWOTRAN - TRACA computer Code (4) to calculate the system K - effective. This code is a version of the TWOTRAN-GG Code (5), modified by the Junta de Energía Nuclear.

A P_0 with diagonal transport correction 4 - group cross - section library was generated using the 69 group cross - section basic library of the WIMS - TRACA code (6). Cell calculations and spatial homogenizing were performed with the WIMS Code (7).

Special attention is being given on the generation of the cross-section library for black absorbers. A detailed spatial calculation using a discrete ordinates method was made for a system composed of a fueled region sourrounded by water and absorber.

A tentative to perform the criticality analysis via XSDRN-KENO IV (8) will be carried out, in order to compare results.

5. THERMAL ANALYSIS

The thermal safety analysis is being carried out by theoretical and experimental heat transfer studies.

This dry storage cask will be designed to remove decay heat passively. Aluminum or aluminum alloys are being considered as basket materials. Cast nodular iron is being considered as containment vessel material.

The exterior cask surface temperature will be less than 82º C, considering an environmental temperature of 38º C.

The heat sources are the decay heat from spent fuel and solar radiation. The cask shall be designed to withstand the fire - test conditions (800º C during 30 minutes).

5.1. THEORETICAL ANALYSIS
(J. Puga and P. Bárcena, ENUSA ; A. Esteban, JEN)

The internal decay heat from spent fuels will be transfered by radiation and conduction to the exterior surface of the cask. Natural convection is not given credit.

The heat transfer from the spent fuel elements to the basket has been only assumed by radiation, according to Watson (9). The basket and the vessel transfer the decay heat radially by conduction. Finally the heat is transfered from the external surface, provided with fins, to the environment by natural convection and radiation.

Several computer codes TAC 3D (10), HEATING 5 (11) etc, are being used to calculate the temperature distribution in the fuel elements, basket and vessel.

5.2. EXPERIMENTAL ANALYSIS
(M. Gispert, M. Montes and A. Chamero, JEN)

The experimental thermal analysis will be carried out in two stages:

1) Simulate a whole fuel element (PWR 17 x 17) in order to check and study axial and radial heat transfer. This element will contain electrically heated tubes and will be full size; the heat generation rate will be of the same magnitude as that of a spent PWR fuel and will be adjusted by the voltage applied to the sub - bundles and the measured resistante of these.

The resistances will be nichrome wire running through ceramic insulators inside the tubes. The temperatures will be recorded by chromel - alumel thermocouples on the in - and outside of the tubes. The heat generation rate will be varied to study different storage conditions and modes. The heat transfer under diferent atmospheres (air and inert gases) will also be studied.

2) Simulate a radial zone of the cask, to check the heat transfer by radiation and conduction. The heat generation rate will be similar to that of a spent fuel assembly and it will be varied to study different storage conditions. The outside temperature of the cask will be varied, in order to study the effect of different environmental conditions.

6. SHIELDING ANALYSIS
(A. Rovira, JEN)

Shield design of the cask involves two radiation sources: gamma rays yielded by fission products and neutrons resulting from (a - n) reactions and spontaneous fissions. Both radiations are to be shielded.

Photons from fission products are divided into groups of energy and employing for each photon group the Kernel attenuation method with lineal build - up and cylindrical geometry.

Neutron (a - n) and neutron from spontaneous fission are analyzed with SABINE-3 (12) using removal diffusion method for neutrons, coupled with calculations corresonding to photons from the activation of materials which are due to these neutrons. These photons are treated with the above mentioned analysis (Kernel attenuation method).

Cilindrical geometry will be used for the calculation or radial shielding; disk geometry source and slab shield on axial direction.

7. STRUCTURAL DESIGN AND ANALYSIS
(M. García Ramírez, ENUSA ; J.J. Rodríguez Ochoa; P. Verón and J. Tovar, ENSA)

The material of the cask has been selected on the basis of its fracture toughness, in accordance with the IAEA regulations which require a ductil behaviour at low temperatures (minus 40º C).

To achieve this, a suitable grade of nodular cast iron has been selected. This material has a great resistance to brittle failure, even at low temperatures, due to graphite nodules which promote plastic deformation around them, impairing the development of fractures by a cleavage mechanism.

This nodular cast iron shows, from the point of view of manufacturing, an excellent castability which allows the moulding of the cask in one piece, in spite of its heavy walls, for shielding purposes, and its thin fins, which allow heat transfer to the environment and act as shock absorbers in case of drop, attenuating the energy of the induced waves.

Considering manufacturing cost, the shorter fabrication time and the lower price of a cast cask, compared to steel, make the nodular cast iron a most atractive material for this use.

Due to the temperature specified for fuel cladding, the choice of the basket material is being addressed to aluminum or aluminum alloys.

The structural analysis will be performed by a 3D base finite element computer program, taking into account mainly the analysis of the 9 m drop test behaviour and the calculation of the sealing surfaces of the two lids.

—————————————————

These as well as other studies and activities have confirmed the feasibility of the basic design concept considered. At the present stage the detailed design activities have been initiated.

BIBLIOGRAPHIC REFERENCES

/ 1 / SEMINAR ON STORAGE OF SPENT FUEL ELEMENTS. OCDE - NEA /JEN June 1978.

/ 2 / REGLAMENTO PARA EL TRANSPORTE SEGURO DE MATERIALES RADIACTIVOS.
Colección Seguridad nº 6. OIEA. Texto revisado (1981).

/ 3 / M. J. Bell "THE ORNL ISOTOPE GENERATION AND DEPLETION CODE" ORNL- 4628.

/ 4 / C. Ahnert and J. M. Aragonés "PROGRAMA TWOTRAN - TRACA".JEN 512 (1981).

/ 5 / K. D. Lathrop, F.W. Brinkley "THEORY AND USE OF THE GENERAL GEOMETRY" TWOTRAM Program. LA - 4432 (1970).

/ 6 / C. Ahnert. Programa WIMS - TRACA. JEN - 461 (1979).

/ 7 / J.R. Askew, F.J. Fayers and P.B. Kemshell "A GENERAL DESCRIPTION OF THE LATTICE CODE WIMS" J. Brit. Nucl. Energ.Society, Oct. 1966 p. 564.

/ 8 / L. M. Petrie and N. F. Cross "AN IMPROVED MONTE CARLO CRITICALITY PROGRAM". ORNL-4938.

/ 9 / J. S. Watson. "HEAT TRANSFER FROM SPENT REACTOR FUELS DURING SHIPPING. A PROPOSED METHOD FOR PREDICTING TEMPERATURE DISTRIBUTION IN FUEL BUNDLES AND COMPARISON WITH EXPERIMENTAL DATA" ORNL - 3439.

/ 10 / S.S. Clark et al. "A GENERAL PURPOSE THREE DIMENSIONAL HEAT TRANSFER COMPUTER CODE". Gulf General Atomic, G.A. 9263.

/ 11 / W.D. Turner. et al. "HEATING - 5. EN IBM 360 HEAT CONDUCTION PROGRAM" ORNL/CSD/TM-15 (1977).

/ 12 / SABINE-3 PROGRAM. EUR 5159.

DISCUSSION

H.J. Wervers, The Netherlands

The dose rate specified for the outer surface of the cask
(40 mr/hr) appears to be rather high in view of the fact that IAEA
regulations require 10 mr/hr at 2 metres from the surface. My
calculations show that for casks of this size the dose rate falls
by up to 50 % from the surface value at a distance of 2 metres.

C. Melches, Spain

The cask will comply with IAEA regulations for transport.
During storage the surface dose is expected to fall between
20 - 40 mrem/hr.

B.R. Teer, United States

Do you have any regulations which limit the dose rate at the
boundary of the site where unrestricted access is allowed ?

C. Melches, Spain

Yes we have; the present regulation for nuclear
installations limits the dose rate to $\leq$ 5mrem/year for unrestricted
areas, which is considered easy to meet.

H. Konvicka, Austria

Do you still follow the wet storage alternative in your
country or have such plans been cancelled more or less ?

C. Melches, Spain

Dry storage is considered to be advantageous, in our
circumstances, compared to wet storage and no more efforts are
devoted to wet storage. Nevertheless, the developments of all
technologies are being followed.

H. Konvicka, Austria

Have you made a comparative cost analysis for wet and dry
storage facilities ?

C. Melches, Spain

Yes we have. For the Spanish need (1500 tU) in the period
1990-2000, wet storage would require more or less the same investment
as cask storage. Cask storage has the advantage in the sense that
investment cost can be staggered and adapted very closely to actual
needs. Cost advantage (total storage cost per t(U)) is in favour of
cask storage.

E.O. Maxwell, United Kingdom

You mentioned that your requirement for storage was 2000 tU, which is a small amount, and therefore you preferred cask storage. Have you done a cost benefit analysis of cask storage versus say, vault storage for this storage capacity. You say cask storage is also very flexible - however, vault storage is also very flexible in that modules can be built to suit operational requirements.

C. Melches, Spain

As I pointed out, our requirements in the period 1990-2000 are 1500 tU. In the present circumstances licensing considerations tend to favour wet storage in basins and dry casks. It is my personal opinion that, in present circumstances, a good temporary solution is cask storage, for the reasons mentioned; its flexibility makes it also easy to substitute by a more appropriate solution and provides adequate time for this. I think that vault storage is certainly a very promising solution.

R. Christ, Federal Republic of Germany

According to our estimates one needs a cavity diameter of about 1500 mm for a capacity of 17 PWR assemblies. On the other hand, you specify a 2000 mm O.D. limit for the cask. That leaves only about 250 mm for both γ and η-shielding. Is that sufficient to meet the 40 mrem/hr dose rate limit ?

C. Melches, Spain

In relation to gamma shielding, calculations show a dose rate of <2 mrem/hr, with a wall thickness of 300 mm. The 2000 mm O.D. specified is not rigid. Neutron shielding requirements depend on the design option to be chosen.

B.Th. Eendebak, The Netherlands

In your criticality calculations you assumed two situations :

1. dry conditions without neutron absorbers;
2. wet conditions with neutron absorbers.

Why didn't you take into consideration the lack of neutron absorber under wet conditions ?

C. Melches, Spain

We will also do this calculation, using the code KENO-IV.

SPENT FUEL DRY-STORAGE SYSTEM FOR TRINO VERCELLESE NUCLEAR POWER PLANT

M. Cuzzaniti, B. Zaffiro, S. Felici, P. Peroni , G. Cane
Ente Nazionale per l'Energia Elettrica
Rome, Italy

ABSTRACT

The paper describes ENEL's design of a dry-storage system for the irradiated
fuel of Trino Vercellese nuclear plant, prepared in 1978 when a critical storage situation was envisaged. The project was carried out as far as obtaining
the Italian Safety Authorities agreement, but was subsequently put aside as
another solution became available.

RESUME

La communication décrit le projet ENEL du système de stockage à sec
pour le combustible irradié de la centrale nucléaire de Trino
Vercellese, conçu en 1978 lorsqu'une situation critique de stockage
était prévue. Le projet a été poursuivi jusqu'à l'obtention de
l'accord par les Autorités Italiennes de Sécurité, mais a ensuite
été abandonné lorsqu'une nouvelle solution fut trouvée.

FOREWORD

In 1978 ENEL, because of the limited fuel storage capacity of its station pools, was faced with the problem of storing the irradiated fuel discharged from the Garigliano (BWR, 160 MWe) and Trino Vercellese (PWR, 270 MWe) nuclear power stations. If prompt alternative storage solutions were not found, very likely the plants would be shut down.

Preliminary negotiations were under way with the European reprocessor, but this solution was considered unreliable because of the uncertainties it entailed. Therefore, ENEL turned its efforts to two approaches that might lead to a satisfactory solution in a short time:

- utilization of the pool of the Avogadro, a decommissioned research reactor; with adequate modifications this pool could accommodate 130 t of fuel;

- construction of a facility of modular concept for dry storage in containers, thus adjustable to the requirements.

Though ENEL preferred the Avogadro pool solution because it would have solved the fuel storage problem up to the end of Trino Vercellese and Garigliano life, the uncertainties associated with the time required for the modifications and issuance of the authorizations made it necessary to carry on the designs and technical-administrative procedures for both solutions, one as standby to the other. The activities continued in 1979 and resulted in the complete design of the dry-storage facility. Subsequently, having speedily obtained the licence for the Avogadro pool, ENEL immediately started the modification work, whereas the dry-storage facility was temporarily set aside. However, with this project ENEL acquired a quite interesting experience because of the new safety aspects it entailed, for which no previous data were available.

DEFINITION OF THE DESIGN

The definition of the dry-storage system was somehow simplified by the fact that the Italian industry was developing the design of a transport container sized to meet Trino Vercellese station requirements. The container, 60-t in weight, with a 35-cm thick steel shield and external neutronic shielding in water, was to be used for transport of irradiated fuel assemblies to the Avogadro deposit. (This container is being manufactured by AGIP Nucleare-Nuovo Pignone on behalf of ENEL and will become available in the second half of 1982). It was thus decided to utilize this transport container design by adapting it to the specific requirements of a dry-storage system. However, ENEL was fully aware that such solution entailed the necessity of unloading the storage containers in the reactor pool and of utilizing proper containers to transport the irradiated fuel to its final destination, once identified.

In view of a four-year decay time of the fuel in the pool, the original design was simplified with regard to heat removal, shielding and nuclear characteristics, and its capacity was increased up to twelve irradiated fuel assemblies.

All these modifications were considered acceptable because the storage containers were going to be housed in concrete structures and the storage area was situated within the boundaries of the Trino Vercellese station, therefore subject to the stringent requirements applied to nuclear plants.

The criteria followed in converting a transport container in a suitable dry-storage container are described below.

1.   Shielding Design

In view of the long decay time of the fuel in the station pool and the consequent reduction in its activity, it was expected that with the same 35-cm thick steel shield the overall external activity (gamma + neutronic) could be kept within acceptable limits without the neutron shield provided for in the original design. Moreover, the container would be housed in an external concrete structure to complete the shielding effect. The design verification was based on conservative assumptions, that is, an average irradiation of 33,000 MWD/MTU (higher by 13% than the actual values), and an initial enrichment of 4% for the evaluation of the neutron dose and of 4.5% for the evaluation of the gamma dose. Furthermore, worsening factors of 20% were introduced for the neutron source and for neutron and gamma attenuations to take into account the uncertainties entailed by the calculation method. Thus, the overall dose intensity (neutron + gamma) was equal to 170 mRem/h on the container and to 40 mRem/h at two meters from the container.

By locating the container in a 0.35-m thick concrete structure, doses equal to 7.4 mRem/h at contact, 2.3 mRem/h at two meters and 1 mRem/h at four meters from the container were obtained.

2.   Thermal Design

After a four-year cooling period in the station pool, the twelve fuel assemblies in each container should have a residual power of 5 kW. Thus, it was deemed possible to keep the maximum temperature on the external container surface within the limit of $80^{\circ}C$ without resorting to finning. This was verified on the basis of conservative assumptions such as an ambient temperature of $38^{\circ}C$, absence of heat exchange by air conductivity and convection in the container (it was assumed that the heat would be transmitted only through irradiation and through the contact surfaces between the internal basket where the twelve fuel assemblies are accommodated and the internal container wall), absence of axial conductivity in the container with heat transferred to the environment only through the lateral surfaces. Furthermore, a peak factor of 1.2 was applied to the axial distribution of the decay power.

The design of the external concrete housing provided for air inlets to ensure an air flow of 1200 kg/h (velocity = 0.2 m/sec).

In these conditions, to improve thermal emissivity it was necessary to design a carbon-steel basket instead of the previously designed stainless-steel basket. Thus, the temperature on the fuel assembly sheath decreased to $324^{\circ}C$, with a corresponding rod temperature of $388^{\circ}C$.

3.   Nuclear Design

From the nuclear design point of view, the worst conditions occur during loading of the assemblies, when the fuel is immersed in water. The nuclear assessment during fuel loading in the pool was carried out by assuming the most conservative conditions, that is, twelve fuel assemblies in the container in a carbon-steel basket without neutron absorbers, an initial enrichment of 4.5% and an irradiation of 20,400 MWD/MTU in fresh water at $20^{\circ}C$. In these conditions, and with the addition of worsening factors such as the fabrication tolerances and the calculation method uncertainties, the calculated k_{eff} was equal to 0.928 and therefore it was deemed possible to utilize a basket without neutron absorbers. This approach was acceptable also because the irradiation of the fuel assemblies in the pool was higher than the assumed value, the water in the pool was borated to no less than 2,600 ppm, and in view of the adoption of very strict control procedures for fuel loading. Visual checks of the identification number of each assembly and checks on the neutron moltiplication factor during loading were specifically required.

For the storage conditions, the evaluated k_{eff} was less than 0.3. It is to be noted that, should conditions different from those considered occur--for instance a lower irradiation or a higher enrichment--the modifications would involve only the basket, which should accommodate less fuel or include neutron-absorber materials. The remaining aspects of the design should not be reviewed because they would be more conservative.

DRY-STORAGE SYSTEM

The dry-storage system requires the construction of modular reinforced concrete structures, where the steel containers--each loaded with twelve irradiated fuel assemblies--are located horizontally. Fig. 1 shows a scale model of six moduli, as many as ENEL expected to build in the first stage.

The storage facility was going to be located in a 40x15 m zone of the Trino Vercellese station area (Fig. 2) that was particularly adequate for its connections with the fuel building and because, being located at the SE end of the station area with the Po river on one side and the radwaste building on the other, it did not interfere with the other parts of the plant.

Each modulus consisted of reinforced concrete components assembled in situ (Fig. 3), that is: the floor and the lower parts of the two lateral walls, which form the main body; the upper parts of the lateral walls; the front and back doors and the top. The floor is encased in the ground and lies on a foundation mat; the two parts of each lateral wall are located by means of dowels. The lateral walls have four cradles to hold the container in its position after removal of the trolley.

Ventilation of each modulus is ensured by natural circulation; the air enters through air inlets protected by grates in the lower part of the two doors, is conveyed towards the center by means of diaphragms and then is discharged through lateral outlets located immediately below the top.

The modulus external dimensions were 5x2.5x3 m, with a wall thickness of 0.35 m.

DESCRIPTION OF THE STORAGE CONTAINER

The storage container, designed by AGIP Nucleare-Nuovo Pignone, consists of a forged hollow carbon-steel cylinder (Fig. 4) with two flanged lids bolted to the cylinder; sealing between the cylinder and the lids is ensured by metal gaskets of the Helicoflex type (a patent of CEA, France).

Seven trunnions on the cylindrical body are used to lift and hold the container in the modulus. Drain valves located on the lids ensure discharge of the water from the container.

The container dimensions are:

- external length   3,900 mm
- internal length   3,300 mm
- external diameter   1,728 mm
- internal diameter   1,068 mm
- weight (with fuel and water)   60 t.

Each container can house twelve fuel assemblies located in as many channels of the 7-mm thick carbon-steel basket . Correct positioning of the twelve channels is ensured by six carbon-steel spacing grids 20 mm thick; four tie rods keep the grids in the right position. The whole system allows regular downflow of the water from the container during drainage.

Once the fuel assemblies have been loaded, the container is transferred to the decontamination cell where, after decontamination of the external surface, it is drained and dried by means of air ventilation. For this operation use is made of the drain valves; in fact, drying is obtained by comecting the container to the ventilation system to let the air circulate from top to bottom, and by checking the moisture content of the air discharged. The container is then left in the cell with a valve open until thermal equilibrium is reached.

HANDLING OF THE CONTAINER

In the fuel building the container is handled with the normal station equipment and following the procedures adopted for transport of irradiated fuel containers to reprocessing plants.

Once the container has been drained and decontaminated, it is located horizontally on an ad-hoc trolley towed by a tractor, and carried to the storage modulus where it is positioned over the trunnion supporting cradles. Then, the trolley loading deck is hydraulically lowered leaving the container in the modulus.

The trolley consists mainly of a box-type steel structure connected to the wheel axis by means of hydraulic jacks that allow lowering of the loading deck.

REQUIREMENTS OF THE CONTROL AUTHORITIES

During the design definition phase ENEL had several meetings with the Italian Control Authority (CNEN). These meetings regarded mainly the calculation methods used and their qualification.

Upon CNEN's request the concrete moduli were modified to make them capable of withstanding external events such as reference tornadoes and related missiles, and the design earthquake for the Trino Vercellese station. The position of the air inlets had to be modified in order to avoid clogging, for instance from snow or other matters.

The radiological conditions of the storage facility were carefully checked with particular regard to the station personnel exposure and to radioactive releases. In particular, CNEN requested a direct check on each container to measure radioactive releases through the lid seals; should the specification values be exceeded, the lids were to be welded to the cylinder. As regards heat removal and in particular the advisability of removing the container fins, the values obtained with the calculation methods were to be verified with experimental tests on prototype.

CNEN also requested to evaluate the consequences of a handling accident and to identify the reliability level requested of the handling equipment, besides an evaluation of the dose to the personnel during loading of the fuel up to location of the containers in the moduli. This evaluation, performed on the basis of previous experiences of irradiated fuel handling and by adding a worsening factor of 20% for unforeseen events, gave an indicative value of the integrated dose absorbed by the personnel lower than 100 mRem/man for each container.

At the end of 1979 ENEL's design, integrated with the aforesaid provisions, was ready for implementation.

ECONOMIC CONSIDERATIONS

The economic evaluation made by ENEL in 1979 on the basis of the container manufacturer quotations is given below.

It should be noted that this evaluation takes into account the fact that the storage facility was located in the area of an existing station and therefore evaluation of the costs for land purchasing, fencing and guard-house, access roads, health physics checks and for the container handling personnel were not necessary. The quotations are given in US dollars 1979 and are obtained by applying the rate of exchange valid at the time.

No. 6 containers	6 x $325,000 =	$ 1,950,000
Civil works and moduli	"	200,000
Trolley	"	60,000
		$ 2,210,000

The unit investment cost was therefore 85 $/kgU.

If the 1979 costs were to be related to date, the aforesaid values would not change greatly; in fact, the inflation rate in the later years is practically counter-balanced by the variation in the dollar exchange rate, which from 1979 to date has been re-evaluated by over 50% in respect of the Lira.

Because of the non-commercial nature of the facility designed by ENEL, its particular limited dimensions and its location within a nuclear station area, we do not deem it possible to compare the aforesaid evaluation with any other evaluation given in the specialized literature.

Finally, the economic evaluations performed lead us to wonder whether the possibility has been fully examined of using free areas of existing stations as temporary storage of not negligible amounts of irradiated fuel previously cooled in the station pools, thus ensuring a greater autonomy to station operation. Such storage systems, that make use of already existing infrastructures, might be an alternative to large centralized storage facilities with advantages not only from the economical standpoint, but also from that of the acceptability by the public opinion.

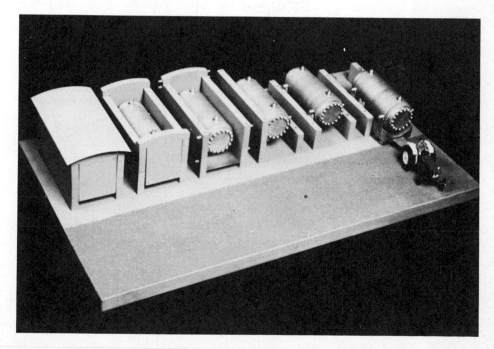

Figure 1   Artistic Impression of the Storage Facility

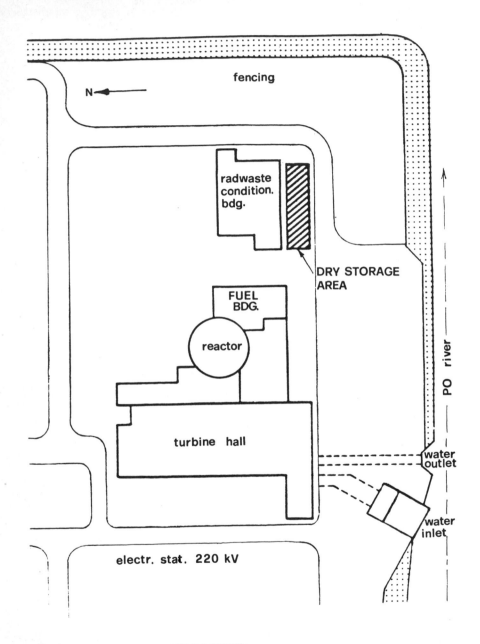

FIG: 2: TRINO PLANT LAYOUT

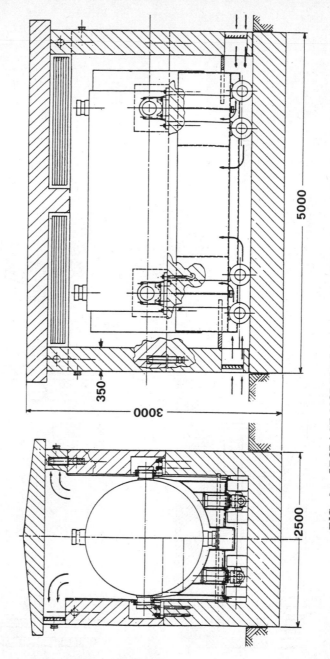

FIG. 3: STORAGE MODULUS

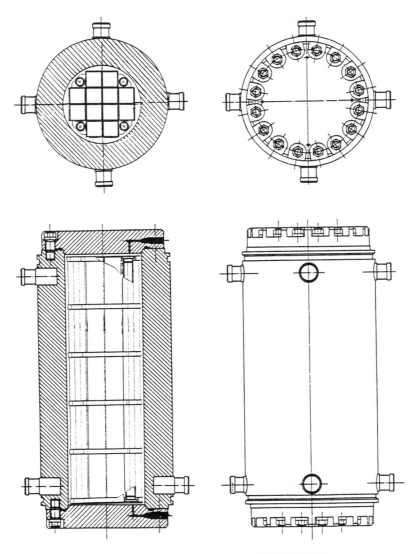

FIG. 4: DRY-STORAGE CONTAINER WITH BASKET

DISCUSSION

C.J. Ospina, Switzerland

There is only one barrier lid in your cask and therefore the question arises about control and long-term operation particularly with regard to cask leakage and how to repair it ?

M. Cuzzaniti, Italy

An agreement was reached with the Italian Safety Authority for a leak control during storage. In any case one must not forget the peculiar situation of our dry storage facility : being in the station area, at a very short distance from the fuel building, a leaking container could always be brought back to the pool, the fuel discharged and the necessary repairs carried out.

A. de Ubieta, Spain

The standards regulating criticality analysis indicate that the said analysis has to be made under optimum moderation conditions which do not necessarily occur with fresh water but rather - theoretically speaking - could arise with a different medium such as mist, water vapour or other. Have you performed an optimum moderation sensitivity study to search for that optimum condition ?

M. Cuzzaniti, Italy

No, the Italian Safety Authority did not require such an analysis in 1979.

G.A. Brown, United Kingdom

A rod temperature of up to 388°C is quoted for storage in air. The general trend now seems to be up to 250°C. Would you agree that your temperatures seem to be too high ?

M. Cuzzaniti, Italy

Yes, I can agree with you, if you refer to today's trend; however, in 1979 the Italian Safety Authorities were satisfied with the results of our calculations.

G.A. Brown, United Kingdom

Do you intend to store unfailed fuel only ?

M. Cuzzaniti, Italy

Yes, the Trino Vercellese fuel is still SS clad and no failed fuel has been recorded up to now.

B. Vriesema, The Netherlands

Did you measure or test the decay heat removal of the cask in a horizontal position ?

J. Fleisch, Federal Republic of Germany

Until now, experimental data on the decay heat removal of a cask in a horizontal position are only available from inactive heater tests. The Würgassen program does not cover any temperature data collection from a horizontally orientated cask. In the Gorleben AFR-facility the casks are arranged in an upright position.

S.J. Naqvi, Canada

Do you agree that your conclusions that fuel cladding rupture is caused first and foremost by creep deformation ignore two important cladding degradation mechanisms :

1. stress-corrosion-cracking (SCC)

2. metal vapour embrittlement (MVE).

Also, if the fuel were defective, do you not have to be concerned about UO_2 oxidation ?

J. Fleisch, Federal Republic of Germany

The cask and its lid system constitute a safe containment even for the hypothetical case of a 100 % pin rupture. The oxidation of UO_2 in case of defective fuel could be completely excluded, because the fuel is stored in an inert gas atmosphere (helium) during the whole storage period and the lid system integrity is controlled continuously by interspace monitoring.

M. Peehs, Federal Republic of Germany

Let me answer the question concerning SCC. To produce SCC you must fulfil three conditions simultaneously. The active fission product must exceed the critical concentration. Hoop stress or strain must exceed a minimum number and the time period must be long enough. According to experimental results all active fission products released during service will have reached chemical equilibrium. At storage temperature no further fission products will be released. Consequently, condition one is not fulfilled. So we can deny the occurrence of SCC. These arguments are also backed up by laboratory tests which show that even when you have enough SCC available the mechanical load typical for storage does not exceed critical numbers.

S.J. Naqvi, Canada

Do you think a test demonstration of 6-10 months is sufficient to predict long-term behaviour ? What would happen when fuel temperatures drop to lower values, to say, 200°C, and cladding hardens ? Do you think under those circumstances, SCC would damage your fuel ?

J. Fleisch, Federal Republic of Germany

The storage period is not the limiting parameter, because following decreasing fuel temperature, creep deformation after a 6-10 month period is a failure mechanism of only minor importance.

SESSION 2

Chairman - Président

M. PEEHS
(Federal Republic of Germany)

SEANCE 2

MODREX : STORAGE IN CONCRETE SILOS AND/OR IRON CASKS

P. Doroszlai, Electrowatt Engineering Services Ltd.

Zurich (Switzerland)

The acronym MODREX stands for Modular Dry Expandable, a concept for the dry storage of LWR spent fuel elements in silo or casks. This double concept offers inherent safety, low exposure to the operating personnel and also significant operational and economic advantage.

The silo storage facility consists of a receiving station for the transfer cask and a series of storage modules. The storage module - the basic element of the facility consists of an individually poured concrete monolithic structure containing nine vertical silo positions dissipating the decay heat of the enclosed fuel elements by a combination of passive convection cooling and heat pipes.

Each silo receives one storage canister which in turn holds several LWR fuel elements or vitrified waste zylinders. It provides a safe and efficient means of handling the fuel.

On the other hand, the MODREX storage cask consists of the same canister as used in silos, loaded in a nodular cast iron cask providing shielding and mechanical protection, as well as heat dissipation through fins.

The combined use of the two dry storage MODREX concepts - silo and cask - at the same or different sites offers substantial operational flexibility. The fuel canister initially loaded in casks can later on be transferred into silos, with the empty cask being reusable at other sites.

MODREX can be considered to consist of three parts, namely: storage canister; storage cask; storage module. A need for extra spent fuel storage capacity stimulated the development of MODREX. The system was therefore designed with this use in mind. During development it was realised that the same design could be used to supply storage capacity for high level waste vitrified blocks. Such usage of MODREX has been prompted by current thinking on the question of high level waste disposal, namely to store the waste for 100 years or more before disposal. The combined MODREX system can play an essential role in a strategy for spent fuel and high level waste management.

Storage Canister

Before describing the design of the MODREX storage canister it is important that its role as the kernel of the MODREX system is explained.

The storage canister is designed to contain the irradiated fuel from the point of loading (at the power plant), during transportation (within a cask), throughout all stages of storage (described below), up to unloading at the re-processing facility. Its function is to provide containment only - other features of the transport cask or storage system provide physical protection. The storage canister concept avoids any direct handling of fuel between the power plant and the reprocessing facility. A similar role applies in the case of vitrified waste blocks.

As the concept and eventual design of MODREX were evolving particular attention was paid to make the system compatible with the existing hardware and practice at nuclear power station and nuclear reprocessing plants. The main aim was that any intermediate storage required between the spent fuel pond at the reactor site and the cooling pond at the reprocessing site could be incorporated with a minimum of change. In achieving this aim, the storage canister has served not only as an integral part of the MODREX system but also as a very flexible interface to existing hardware and practice.

The MODREX storage canister, which is made of stainless steel, has a lid and a three part body; a curved or dished bottom section, an annular main body and a flange. The lid is composed of a ring flange and curved plate. It is equipped with valves and handling fittings to facilitate loading and wet-to-dry conversion in an existing fuel loading pool. A drop accident of 7 metres, the maximum drop possible during operations, can be withstood by the dished bottom section without loss of containment. On the outside of the main body there are longitudinal fins which assist heat transfer, locate the storage canister centrally either in a transport cask or in the MODREX storage cask or MODREX storage module and provide protection against impact. The main function of the storage canister is containment and the thin walls do not provide radiation protection. This storage canister is a unique feature of MODREX.

There are two versions of the storage cask devolopped. External dimensioning of the first for seven PWR fuel elements has been such that it can fit within a modern large transport cask for shipping twelve unconsolidated PWR fuel elements and can also be fitted within both the MODREX

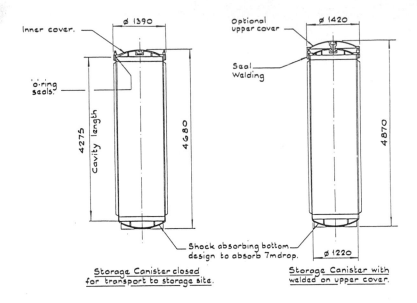

Inner cover.

o·ring seals.

Cavity length

4275

ø 1390

4680

Optional upper cover

Seal Welding

ø 1420

4870

Shock absorbing bottom design to absorb 7m drop.

ø 1220

Storage Canister closed for transport to storage site.

Storage Canister with welded on upper cover.

Figure 1

Storage Canister with Various Baskets.

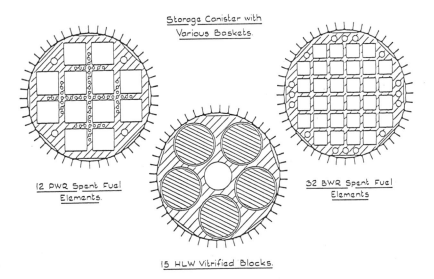

12 PWR Spent Fuel Elements.

15 HLW Vitrified Blocks.

32 BWR Spent Fuel Elements

Figure 2

storage cask and MODREX storage module silo. A larger version
of the canister developped recently may take twelve PWR fuel
elements. It fits in larger transport-, resp. storage casks
and also in corresponding silos.

A sealed, dry, inert (helium) atmosphere is maintained
inside the storage canister when loaded. For long term
storage a second upper cover can be welded on as illustrated
in figure 1. The helium atmosphere assists the heat transfer,
avoids corrosion of the fuel cladding and serves as a leak
detection sensor. Internally the load is secured by an
aluminium basket. There is a different basket design for
each load type. Basically the basket has channels to locate
the fuel or waste and holes for criticality control under
wet loading conditions. To date three designs have been
developped: (i) PWR spent fuel for 12 unconsolidated elements
or 23 consolidated elements (where consolidation factor is
1.9); (ii) BWR spent fuel for 32 unconsolidated elements or
61 consolidated elements; (iii) high level waste vitrified
blocks 15 short (1.30 m) blocks or 10 long (1.85 m) blocks.
Cross-sections for these three designs are presented in
figure 2. The basket is designed to allow for thermal
expansion. It is removable so that the same storage canister
can be used for different load types.

The loading procedure for spent fuel at the fuel pond is
modified only in that the storage canister is firstly placed
within the transfer cask and has to be sealed prior to
sealing the transfer cask. To all intent and purposes it
performs the function of a loading basket during the operation.
Direct handling operations of the fuel have thus been kept
at two, namely loading once and unloading once, and thus
reducing the probability of an accident were additional,
intermediate handling operations necessary. Once the storage
canister is sealed, it is transported as a single unit and
the seal does not need to be breached until the spent fuel
is required for reprocessing. The unloading procedure at the
reprocessing plant cooling pond has similar minor adjustments to the loading procedure.

Storage cask

The MODREX storage cask because of its relatively high price
is seen as a short term solution whilst the utility decides
on its longer term storage requirements. It also acts as a
buffer storage during the construction and licensing of a
more economical permanent construction. The storage cask has
the ability of being transported empty from one site to
another as a heavy load and not as a radioactive load.
However in countries where nodular cast iron may be licenced
for transport it may be used for overland transport of the
loaded canister as well, thus replacing the forged transfer cask.

As this storage cask is aimed at short-term and buffer
storage, it is designed to be compatible with the eventual
long term storage option. For this reason the MODREX storage
cask has been designed basically around the storage canister.
It is therefore compatible with both the MODREX storage
canister and the longer-term MODREX storage module. Electrowatt Engineering Services has been involved in the design of
casks of various sizes for the transport of spent fuel.
Bases on this considerable design experience the MODREX
storage cask evolved, suitable for spent fuel storage
applications.

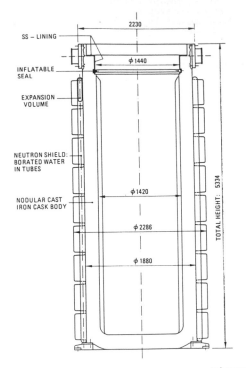

Figure 3

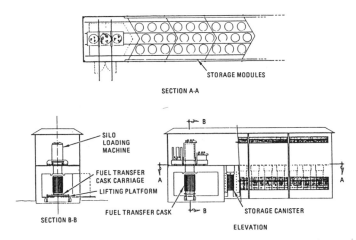

Figure 4

The MODREX storage cask provides a storage unit for a single MODREX storage canister. The basic design has double independent containment - the canister is sealed and the cask is sealed.

A cross-sectional view of the MODREX storage cask is presented in figure 3. Made of nodular cast iron, the storage cask is a massive cylindrical container which provides both radiation and mechanical protection. The cylindrical portion of the cask has external longitudinal fins to dissipate the decay heat from the load. The spaces between these fins contain steel pipes filled with the neutron shielding material.

In circumstances where it is necessary to load spent fuel into the MODREX storage canister within the storage cask, a protective skirt of stainless steel has been designed to permit this optperation in the fuel pond at the reactor and avoid contamination of surfaces.

Storage module facility

The MODREX storage module facility consists of a transfer cask receiving area beneath a loading bay, a movable loading machine, a number of storage modules each with nine silo positions and a weather protection building. These constituent parts are indicated in figure 4.

Modularity and expandability, essential characteristics of MODREX, are provided by the storage module. This consists of an individually poured chevron shaped concrete block with nine silo positions. It has a system of interconnected air ducts and heat pipes which provide indirect cooling of the storage canister by natural circulation of air. The storage module is illustrated in figure 5.

When the storage canisters are not being moved, the only task for the operators is the occasional controlling of the monitoring systems. The number of personnel required, therefore, is minimal.

The design lifetime for the MODREX storage module facility is 120 years, as a minimum.

Cooling system: At the base of each silo six ducts connect the air space in the silo to a common air space below the silos at bottom of the module. This common air space can also provide access for cleaning etc. before the silos are loaded. Once the silos are loaded this area will become a radiation zone. The weight of the storage canister is taken on a concrete pedestal which rises through the common airspace to form the base of the silo.

The air in the common airspace underneath the silos is relatively cold. It rises up the four ventilation ducts into the silo, where it cools the storage canister and the silo lining, and up to a plenum. Two of the ducts allow the air into horizontal channels which permits the mixing of warm air between silos and between modules. The remaining four horizontal ducts lead the hot air in the plenum to four separate down-draft channels. In these channels the air gives up its heat to the heat pipes and continues, relatively cold to the common airspace at the bottom each of the four down-draft channels is cooled by a different set of

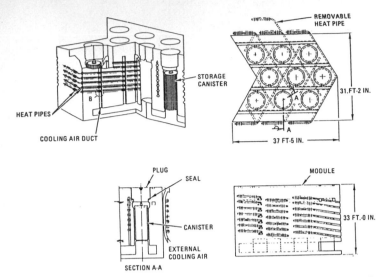

STORAGE CANISTER

HEAT PIPES

COOLING AIR DUCT

B

REMOVABLE HEAT PIPE

31 FT-2 IN.

37 FT-5 IN.

A

PLUG

SEAL

CANISTER

EXTERNAL COOLING AIR

SECTION A-A

MODULE

33 FT.-0 IN.

Figure 5

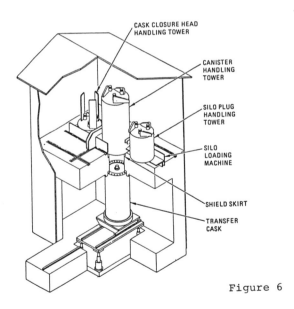

CASK CLOSURE HEAD HANDLING TOWER

CANISTER HANDLING TOWER

SILO PLUG HANDLING TOWER

SILO LOADING MACHINE

SHIELD SKIRT

TRANSFER CASK

Figure 6

heat pipes. This factor and the free mixing of air at the top and bottom of the blocks, ensure that if any failure in the heat pipes or blockage of the air ducts occurs, excess heat loading is shared by all the air ducts and heat pipes surrounding the affected area, and so the resultant increase in temperatures is minimised. This is the multiple redundant inner circuit for heat transfer.

The heat pipes serve as a transfer circuit and are standard engineering components, consisting of long tubes containing a liquid in equilibrium with its vapour. Suitable liquids which have their vapour pressure close to atmospheric at the temperature range of interest are methanol, ammonia or water. Choice of working fluid may be determined by external ambient temperatures at the selected site. The heat pipes are installed at a shallow angle in separate channels cast into the concrete, the heated end being lower than the cooled end. In the section crossing the down-draft channels the heat pipes are finned to improve heat transfer. The number of heat pipes required for each down-draft channel will depend on the liquid selected and the heat load to be handled. Up to eight can be fitted one above the other in the position illustrated. Eventually as the heat source decays these heat pipes are no longer be needed.

On the outside faces of the module the heat pipes are also finned and heat is lost to the atmosphere. The air is led in from the outside through louvres, rises past the heat pipes where it is warmed, past the inner and outer skin of the weather protection superstructure and out via a second set of louvres. This is the external circuit for heat transfer and avoids the need for any obvious chimneys.

Heat removal is designed in such a way that the concrete need only conduct the heat generated within itself and its temperature never exceeds $80^{\circ}C$ for a heat load of 12 kW per storage canister.

Loading machine: The only active operations required are related to the transfer of the MODREX storage canister to and from the silo locations. These transfer operations are performed with the use of a movable loading machine. This loading machine handles the lid of the transfer cask, the storage canister and the concrete plug of the silo. Since the storage canister has to be protected and shielded during the transfer operation, the loading machine consists of protecting towers for the cask lid, the storage canister and the concrete plug.

The original loading machine designed in 1978/79 for storage module operation is shown in figure 6 where it is located in the loading bay on top of the receiving station in which a transfer cask is located. The three protective towers can be clearly seen. This design could not, however, withstand the impact of an aeroplane even though the storage modules and the receiving bay could. This aeroplane impact resistance of the loading machine is a requirement from some national licensing bodies, e.g. Germany and Switzerland. A modification to the loading machine was considered in order to comply with this requirement rather than attempt to design the weather proof superstructure to be aeroplane impact proof. The major problem is damage to the storage canister during emplacement to, or withdrawal from the silo. The storage canister protection tower was the part of the loading machine modified. An internal annulus of forged steel was

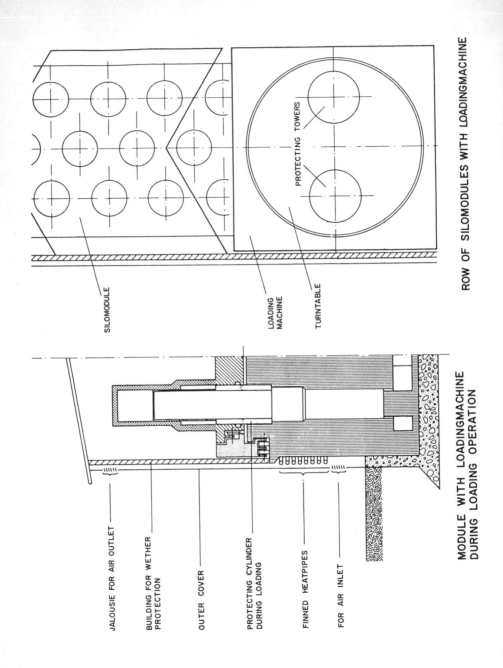

ROW OF SILOMODULES WITH LOADINGMACHINE

SILOMODULE

LOADING MACHINE

TURNTABLE

PROTECTING TOWERS

MODULE WITH LOADINGMACHINE
DURING LOADING OPERATION

JALOUSIE FOR AIR OUTLET

BUILDING FOR WETHER PROTECTION

OUTER COVER

PROTECTING CYLINDER DURING LOADING

FINNED HEATPIPES

FOR AIR INLET

Figure 7

- 60 -

included. This annulus would be lowered into the silo and
rest on the plug ledge. When the machine and the annulus are
fixed in position the storage canister can then be lowered
or raised and is protected from damage due to aeroplane
impact. In additional, the carriage of the transfer machine
is changed the way it interlocks with the top of the siloblocks.
The protecting towers are safely secured to a turntable on
the top of a carriage, similar to the original design back
in 1978. The principle of this new machine is shown in
figure 7.

Tightness control system

The loaded storage canisters are sealed either mechanically,
or, for long term storage, with welded cover.

The tightness of this seal, which is the most important
barrier between the stored radioactive material and the outside air, must be regularly verified. The method proposed
for this verification is based on monitoring the pressure of
the helium inside each storage canister.

The storage canisters are filled with helium during their
loading operation so that the internal pressure will be 1.3
to 1.5 atmospheres when the canisters is in the concrete
silo under normal operating conditions. The high volatility
of the helium guarantees that it will leak out through the
smallest failure in the sealed canister long before any
gaseous or aerosol fission products could do so. Therefore,
the pressure monitoring system to be described will warn of
a containment problem long before any fission product release
could occur. Canisters with very low leakage rates of helium
(for example, 0.1 atmosphere over a few years) may be considered to be absolutely tight as far as fission product release is concerned.

In addition to guaranteeing no release of activity, the
pressure monitoring system can be used to verify the
presence of helium inside the canister. This being the case,
the physical properties of helium may be considered in
determining the heat transfer mechanism within the canister.

Because of the high volatility of helium, the method for
monitoring the pressure inside the canister will not use any
penetration of the canister for cables or tubes. Instead,
the method is based on measuring the expansion of a very
high-quality seal-welded bellows filled with helium which is
mounted inside the canister (see figure 8). One end of the
bellows is fixed while the other end is mechanically connected
to a piece of mild iron bar so that any displacement of the
free end of the bellows results in an amplified movement of
the iron bar. The iron bar is plated with stainless steel to
avoid corrosion resulting from the wet loading/unloading
operations.

Any pressure drop inside the canister will result in an
expansion of sealed bellows, which in turn will move the
position of the iron bar. The position of this bar can be
read from óutside the canister by a row of high-frequency
coils mounted in the storage silo just beneath the canister.
Since the stainless steel canister wall is permeable for
magnetism, this method is not influenced by the thickness of
the wall.

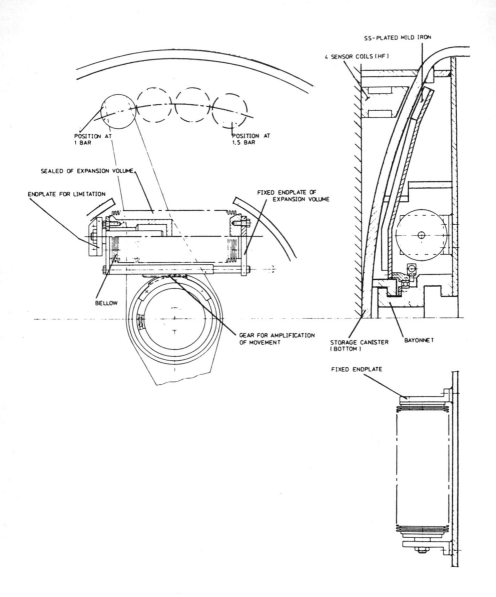

POSITION AT
1 BAR

POSITION AT
1,5 BAR

SEALED OF EXPANSION VOLUME

ENDPLATE FOR LIMITATION

FIXED ENDPLATE OF
EXPANSION VOLUME

BELLOW

GEAR FOR AMPLIFICATION
OF MOVEMENT

SS-PLATED MILD IRON

4 SENSOR COILS (HF)

STORAGE CANISTER
(BOTTOM)

BAYONNET

FIXED ENDPLATE

pressure checking device   Figure 8

For periodicly energyzing the high frequency coils as well
as for temperature and radiation monitoring of the internal
air circuit a microcomputer system is provided. Thus the
pressure in each individual storage canister is recorded and
may be analyzed. Any sudden change in the evolution of this
pressure will indicate the need for immediate inspection for
cause. In addition, the records may be used for inventory
control. Thus the stored active waste is continously accounted
for.

The air inside the blocks is not completely isolated from the
atmosphere, as the sealings at the heat pipes penetrations of
the siloblock's wall as well as the sealings below the silo
plugs are not considered to be absolutely airthight. Because
of changing meteorological conditions and during the loading
procedure a periodical air exchange between the inside air
and the atmosphere can not be excluded. A small filtered air
extraction system, however ensures that any leakage occurs
invard, thus the silo block may be considered as a second
barriere for the relase of fission products.

MODREX which combines the use of a storage with both a
concrete silo facility and storage casks has various unique
features. Due to the short lead time for licensing of the
storage cask, the quick availability of storage capacity is
assured. In the meantime the construction of the storage
silo can proceed and eventually storage canisters can be
transferred from storage casks to the concrete silo. This
approach has a substantial financial advantage, since the
concrete silo is more economical than the cast iron casks.
The empty storage cask may be re-used at other sites.

Summary

As with other dry storage facility designs MODREX has low
maintenance requirements, minimal secondary waste arisings,
long storage life, little or no corrosion to fuel cladding
and is suitable for both spent fuel and high level waste
usage.

The unique features incorporated into MODREX to improve
safety and to decrease cost are:

-   containerisation at the fuel loading by of the power
    station.

-   the availability of double containment (a requirement in
    some countries).

-   quasi continous tightness control of the containments.

-   an indirect passive cooling system.

-   the availability of an aeroplane impact proof operation.

-   early warning leakage monitoring system.

-   low initial capital investment.

-   modularity permiting expansion to meet requirements.

DISCUSSION

H.J. Wervers, The Netherlands

In the drawing of a transport flask an inflatable rubber seal was shown. Apparently this is within the radiation zone. In view of this will the rubber have a reasonable service lifetime ?

P.G.K. Doroszlai, Switzerland

The inflatable rubber seal is activated only during wet loading/unloading of the storage cask and canister, in order to avoid contamination of the space between them. Once the storage cask is closed, the rubber seal has no further function. When the canister is transferred to the concrete silo, the seal of the empty cask may be replaced before further use of the cask.

I. Oyarzabal, Spain

Would it be possible to know some technical details about your concept, like :

- maximum fuel cladding temperatures
- basket material
- heat transfer mechanisms
- heat load.

P.G.K. Doroszlai, Switzerland

The maximum allowable fuel cladding temperature as well as the heat transfer mechanism depend on the credibility of the presence of helium in the storage canister. If the helium may be credited for heat transfer, the allowable cladding temperature is 380°C; radiation and partially convection are the basic mechanism of heat transfer. The expensive aluminium basket may be avoided in this case. If helium may not be credited, the presence of air has to be considered, the temperature limit of the cladding comes down to 250°C, and the aluminium basket with its heat conduction capability becomes necessary. Also the role of convection is more important in this case. The permissible heat load per canister depends on these factors and is between 12 and 15 kW.

The credibility of helium for heat transfer and corrosion depends on the standpoint of the licensing authority involved. In the MODREX canister there is a helium checking device provided for quasi continuous control of the presence of helium.

TECHNICAL AND ECONOMIC ASSESSMENT OF ALTERNATIVE DRY STORAGE METHODS

E. R. Johnson and J. A. McBride
E. R. Johnson Associates, Inc.
Reston, Virginia, USA

ABSTRACT

The paper presents the results of an assessment of four alternative methods of dry storage of spent nuclear fuel in respect to the state of technology, licensability, implementation schedule and costs when the storage is used at a reactor location to supplement existing pool storage facilities. The methods of storage considered were storage in casks, drywells, concrete silos and air-cooled vaults. The impact of disassembly of spent fuel and storage of consolidated fuel rods was also determined. The economic assessments were based on the current projected storage requirements of a U.S. utility operating twin 824 MWe pressurized water reactors. Costs were estimated for a number of combinations of storage mode and packaging processes and considered storage of both intact assemblies and consolidated rods.

1.0   INTRODUCTION

Up until about 1975 it was planned in the U.S. that spent nuclear fuel would be stored at the nuclear power plant site for only a few months, after which it would be transported to a reprocessing plant for recovery of contained uranium and plutonium values.   Accordingly, nuclear power stations designed and built prior to that time generally provided storage space for one to two batch discharges of spent fuel plus the capacity to receive a full core.   Delays in the construction and licensing of reprocessing plants required, however, that some additional storage space be made available. Moreover, the promulgation in 1977 of a national policy indefinitely deferring commercial reprocessing activities [1] increased the motivation for reactor owners to provide additional spent fuel storage capacity at reactor locations.   The most economical method of obtaining added storage capacity, at least for the initial increments, was to re-rack the existing fuel storage pools at power plant sites.   This has been the principal method used to date to obtain needed storage space.   Most new nuclear power plants, i.e., those in the initial stages of construction or in the design and planning stage, have been designed with a capability for storage of as much as 10-15 years discharge of spent fuel, and in a few cases, with life-of-plant storage capability.

The extent to which re-racking can relieve the problem for the older plants is, however, limited by the space available, by the extent to which storage racks can be compacted, and, most importantly for some of the plants, by structural limitations imposed by spent fuel pool design and construction.   Thus operators of these plants will face increasingly pressing requirements to obtain additional storage space for spent fuel. Although the national policy for deferring spent fuel reprocessing is no longer in force, there is little prospect that commercial reprocessing services will be available in the U.S. before the 1989-90 period; several U.S. nuclear power reactors will lose their full core reserve capability in the 1985-89 time period [2].

As a result of this situation, it is now necessary that utility companies operating nuclear power reactors not only utilize their existing water pool storage facilities to the maximum extent possible, but also plan to provide separate storage facilities for any additional storage requirements projected to arise between the time the existing pools are filled and the likely date of repository or reprocessing availability.   The repository is currently expected to be available in the 1997-2006 time frame [3].

There are two basic methods available for storage of spent fuel -- water pool storage and dry storage; the latter may be provided in more than one way: by metal storage casks, in drywells (or caissons) beneath grade, in concrete storage silos, or in air-cooled storage vaults.   Water pool storage of spent nuclear fuel has been used successfully in the United States and elsewhere for over 30 years both for the storage of spent fuel at the reactor site and for the storage of spent fuel at reprocessing facilities.   In fact, water pool facilities must be used by utilities to receive fuel which is freshly discharged from light-water reactors because of the initial high thermal output of this fuel. Dry storage techniques, though not new, have not been employed extensively [4].   U.S. experience dates back to the late 1960's [5-11]; Canadian workers have been operating experimental dry storage units since 1975 [12-14].   Dry vault storage is being used on a full scale basis in England [4,15].

This study was directed toward the development of an assessment of the technology, licensability, implementation schedules, and costs for the possible alternative methods of dry storage of spent nuclear fuel when used to supplement existing water pool storage facilities at reactor facilities. The alternatives which were evaluated were:

(1) storage casks (which may or may not be licensed for transportation over the road)
(2) drywells beneath grade
(3) concrete storage silos
(4) air-cooled vaults

Factors taken into account in the technological assessment included the state of development and demonstration, and the requirements for and availability of needed equipment. In order to provide a broader basis of comparison among the alternatives, similar assessments were made of the use of additional water pool facilities at the reactor site to meet added storage requirements.

2.0 ASSESSMENT OF ALTERNATIVES

2.1 Bases

Each of the dry spent fuel storage methods was evaluated under several combinations of operational options. Cases were examined in which spent fuel assemblies are stored in an as-discharged condition (unconsolidated spent fuel), and where the assemblies are disassembled and the fuel rods stored (consolidated spent fuel). Moreover, each of the spent fuel storage methods was evaluated for instances where the canning and/or fuel rod consolidation operations are accomplished in the reactor storage pool as well as where such are accomplished in separate canning and fuel rod consolidation facilities located at the reactor site. A total of sixteen different cases were examined and evaluated, and a cost estimate was provided for each.

For the purposes of this study, it was assumed that unconsolidated spent fuel stored in drywells, silos and vaults would be canned; and that consolidated spent fuel would be canned for all modes of storage; no canning would be involved for storage of unconsolidated fuel in a water pool or in storage casks. Canning of spent fuel may not necessarily be a regulatory requirement for any of the methods of storage studied, however, canning does appear to be desirable in many instances because of operational considerations. Canning of consolidated spent fuel is desirable and cost effective for all modes of storage for ease of handling of the disassembled fuel rods; canning of both unconsolidated and consolidated spent fuel is desirable and cost effective for storage in drywells, silos and vaults to eliminate the prospects for contamination of the storage system and to minimize the costs for decontamination and decommissioning of the storage facilities at the end of their useful life.

The specific configurations employed as the bases for the economic evaluations are set forth in Figures 1, 2, and 3 for the drywell, site, and vault storage modes; Table I presents a summary of the pertinent characteristics of the casks which were considered for the cask storage option; cask costs were assumed to be $600,000 each for the unconsolidated fuel and $750,000 for the consolidated fuel.

2.2 Methodology

The technical evaluation compared the technical status and suitability of each of the spent fuel storage methods. A review was made of available information regarding each of the storage methods involved; and a system for numerical rating of the alternative storage methods was developed, using an evaluation matrix which considered the following factors:

o Status of technology
o Status of equipment development and demonstration

TABLE I - STORAGE CASK DESIGN DATA

Designer/Manufacturer	REA (1)	Transnuklear (2)	GNS (3)
Model	REA-2033	TN-2100	CASTOR-V(B&C)
Capacity - PWR Assemblies	24	21	20-24
BWR Assemblies	52	37	50-52
Weight, Loaded, Tons	87.5-97.5	110-120	100-125
Age of Fuel, years	5	5-8	5
Thermal Load, kW	30[a]	15	45-55

_____

(a) Can be increased to 47kW by addition of special fins at the storage site.

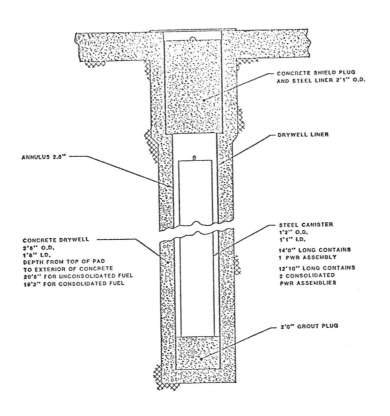

CONCRETE SHIELD PLUG
AND STEEL LINER 2'1" O.D.

DRYWELL LINER

ANNULUS 2.5"

STEEL CANISTER
1'2" O.D.
1'1" I.D.

14'0" LONG CONTAINS
1 PWR ASSEMBLY

12'10" LONG CONTAINS
2 CONSOLIDATED
PWR ASSEMBLIES

CONCRETE DRYWELL
2'8" O.D.
1'8" I.D.
DEPTH FROM TOP OF PAD
TO EXTERIOR OF CONCRETE
20'8" FOR UNCONSOLIDATED FUEL
19'2" FOR CONSOLIDATED FUEL

2'0" GROUT PLUG

FIGURE 1

DRYWELL STORAGE ARRANGEMENT

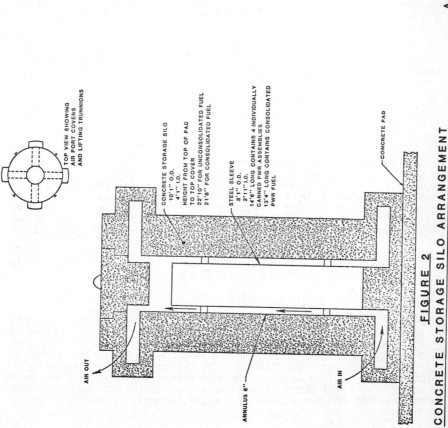

FIGURE 3

AIR-COOLED VAULT STORAGE FACILITY

AIR INTAKE PLENUM

AIR EXHAUST PLENUM

114'0"

16 CANISTERS
2'0" CENTERS
(TYP)

EL 47'6"

AIR FLOW

EL 25'0"

EL 6'0"

EL 0'0"

EL −31'0"

TOP VIEW SHOWING
AIR PORT COVERS
AND LIFTING TRUNNIONS

CONCRETE STORAGE SILO
10'1" O.D.
4'1" I.D.
HEIGHT FROM TOP OF PAD
TO TOP COVER
22'10" FOR UNCONSOLIDATED FUEL
21'8" FOR CONSOLIDATED FUEL

STEEL SLEEVE
3'1" O.D.
2'11" I.D.
14'6" LONG CONTAINS 4 INDIVIDUALLY
CANNED PWR ASSEMBLIES
13'4" LONG CONTAINS CONSOLIDATED
PWR FUEL

CONCRETE PAD

AIR OUT

ANNULUS 6"

AIR IN

FIGURE 2

CONCRETE STORAGE SILO ARRANGEMENT

o   Amount of prior experience with method
o   Operation complexity
o   Quality assurance
o   Retrievability
o   Safety and safeguards considerations

Each factor was assigned a weight representing its relative importance to
the others and each storage method was rated on a scale of 0-10 for each
factor. The weighted grades for all factors were summed to give a figure of
merit for each storage method evaluated. The figures of merit thus
determined were used to establish the relative technological ranking of
each storage method.

The assessment of the relative ease of licensing the different
storage facilities consisted of a review of the licensing-related aspects
of each of the storage methods which considered such factors as:

o   Required reactor license modifications
o   Barriers to the release of radionuclides
o   Active versus passive heat transfer systems
o   Factors impacting the integrity of barriers and shielding
    during the storage period (seismic, corrosion, etc.)
o   Availability of transport medium for released radionuclides
o   Personnel radiation exposure

Based in part on the results of the licensing and technical
assessments, estimates of the time required for design, licensing,
construction and startup of facilities were made for each of the dry storage
alternatives.

Lifetime capital and operating costs were determined for each of
the storage alternatives, and comparisons then made among them on the basis
of
                    capital cost,
                    operating cost,
                    total life-of-facility cost,
                    discounted total life-of-facility cost, and
                    unit storage cost.

The schedules of fuel quantities
and availabilities which were used in the construction of the cost estimates
were those dictated by the currently projected storage requirements of the
Surry Station of Virginia Electric and Power Compay (Vepco); the Surry
station consists of two 824 MWe pressurized water reactors. Surry Unit 1
commenced operation in 1972 and Surry Unit 2 commenced operation in 1973.
The storage capacity of the spent fuel pool at the Surry Station was
increased to 1044 assemblies in 1978 by re-racking, with the resulting net
capacity of the pool being 887 assemblies, plus space for a full core
discharge of 157 assemblies. Total fuel for which supplemental storage will
be required over the remaining lifetime of the Surry units is 2119
assemblies, equivalent to nearly 1000 tons of uranium.

2.3 Technical assessment

The technical assessment employed a semi-quantiative procedure
based on consideration of the seven basic attributes identified in Section
2.2. Each factor was examined and identified and some were further broken
out in lower levels of definition; the weighting assigned to the principal
factor in such cases was distributed to the sub-factors, in proportion to
their relative importance. Each alternative storage method was then
compared with the others in respect to each of the attributes and sub-
factors, and assigned a level of comparative merit, based on a generic

rating scale. The relative weighting of the principal factors and the generic rating scale are set out in Tables II and III, respectively. The summation of the products of the weightings and the individual process rating gave for each of the alternative storage processes a figure of merit. Table IV lists the sixteen alternatives in decreasing order of preference. Although the rating of all the consolidated rod storage options showed them to be technologically less attractive than any of the storage options for unconsolidated fuel, this low ranking is a result almost entirely of the lack of experience with the technology and the lack of adequate heat transfer data. Thus, it is entirely possible that a reasonable amount of experimental work with irradiated fuel rods would result in data which would relieve the uncertainties in these areas.

TABLE II - EVALUATION MATRIX

Status of Technology	2.2
Status of Equipment Development and Demonstration	1.0
Amount of Prior Experience	0.8
Operation Complexity	2.4
Quality Assurance	0.4
Retrievability	1.2
Safety and Safeguards Considerations	2.0
Total	10.0

TABLE III - GENERIC LEVEL OF MERIT VALUES FOR DRY STORAGE PROCESS CHARACTERISTICS

0 - 2   Process involves multiple operational steps, generates significant collateral problems, and involves one or more uncertainties requiring resolution. Minimal experience and data available.

3 - 4   Slightly above average in operational problems, some experience and data available. Some uncertainty.

5   Process presents average operational problems, both direct and collateral. Development or small scale operational data and experience available.

6 - 7   Somewhat better than average in performance characteristics. Limited operational data.

8 - 10   Process is easily managed, generates few collateral problems, and is based on large body of data and experience.

TABLE IV - OVERALL RANKING OF STORAGE OPTIONS IN
TECHNOLOGICAL ASSESSMENT

Unconsolidated Fuel	Ranking	Consolidated Fuel
Drywells, Pool Canned	(1)	
Drywell, Canned in Separate Facility	(2)	
Air Cooled Vault, Pool Canned	(2)	
Storage Casks	(3)	
Water Pool	(4)	
Concrete Silos, Pool Canned	(5)	
Air Cooled Vault, Canned in Separate Facility	(6)	
Concrete Silos, Canned in Separate Facility	(7)	
	(8)	Drywells, Pool Canned
	(9)	Concrete Silos, Pool Canned
	(10)	Water Pool
	(11)	Storage Casks
	(12)	Drywells, Canned in Separate Facility
	(13)	Air Cooled Vaults, Pool Canned
	(14)	Concrete Silos, Canned in Separate Facility
	(15)	Air Cooled Vaults, Canned in Separate Facility

2.4 Licensing assessment

The assessment of licensability of the storage options is a substantially more subjective evaluation than was the case with the technological assessment. It involves primarily questions of licensing time and availability of supporting data; secondary questions relate to the level of uncertainty in the quality of supporting documentation and to the cost of the development work required to produce supporting documentation acceptable to the licensing authority as a basis for authorizing the activity. The comparisons of the alternatives were therefore based on such considerations.

Unfortunately, there are few benchmarks one can use to assess the reasonableness of projections in these areas. Recent experience with the U.S. regulatory authorities does not provide much useful guidance, as the only licensing case under the regulation pertinent to independent fuel storage installations (10CFR72) which has come before the NRC was a request for renewal of the existing "storage only" license at the General Electric Morris Operation; no precedent exists for a full scale licensing action involving a new facility. In summary, the two factors -- time and additional development work required -- will be determined by

o the procedural requirements of the regulations, which require of the order of 28-36 months to complete [15],
o the amount and quality of supporting data available when the application is submitted, which will determine the additional work required,
o the extent of prior licensing experience with the method, which at present consists of the renewal case, and
o the extent of intervention in the license hearing proceeding.

Based on these considerations, the following is believed to be the probable order of increasing difficulty of licensing of the ten basic storage processes:

(1) Water Pool Storage, Unconsolidated Fuel
(2) Drywell Storage, Unconsolidated Fuel
(3) Cask Storage, Unconsolidated Fuel
(4) Concrete Silo Storage, Unconsolidated Fuel
(5) Water Pool Storage, Consolidated Fuel Rods
(6) Air-Cooled Vault Storage, Unconsolidated Fuel
(7) Drywell Storage, Consolidated Fuel Rods
(8) Concrete Silo Storage, Consolidated Fuel Rods
(9) Cask Storage, Consolidated Fuel Rods
(10) Air-Cooled Vault Storage, Consolidated Fuel Rods

2.5 Economic assessment

The economic assessment was constrained by the requirement that it be based on the supplemental storage needs of a specific utility, and in particular by the fact that the utility required supplemental storage beginning at an earlier time than several of the alternatives could be available. It was assumed in these cases that the storage requirements between the time of initial need and the time of availability of the particular alternative installation would be met by the use of storage casks.

For purposes of arriving at uniform estimates of elements common to two or more of the sixteen different cases, the cost estimates were broken out into a series of cost modules, in each of which capital and operating costs were determined; in the cases of the modular storage concepts -- the cask, drywell, and concrete silo options -- the costs were further divided into the costs of the storage units themselves and the costs of the supporting storage facilities required. By appropriately combining these cost modules, the total cost of any of the sixteen options was obtained.

3.0 CONCLUSION

The overall results of the comparative assessments of the spent fuel storage alternatives is presented in Table V. It is to be emphasized that these are preliminary conclusions, based on presently available information, and based on the storage needs of a single, two-unit nuclear power station. An assessment based on the requirement to provide a large away-from-reactor storage capacity would quite likely lead to different conclusions. Elimination of the constraints on the time schedule for providing incremental capacity, together with optimization of the design of facilites, would quite likely result in a decrease in the estimated costs.

Within these limitations, this preliminary assessment of the alternative methods available for use in providing supplemental spent fuel storage capacity at reactor sites leads to the following conclusions:

3.1 The dry storage of consolidated spent fuel in passive storage systems should not be seriously considered for general use by utility companies at the present time. Additional research, development and demonstration work is necessary concerning (a) heat transfer characteristics and maximum fuel clad temperatures of consolidated fuel during such storage, (b) disassembly of spent fuel in the reactor pool and the containment required for any loss of integrity of fuel during disassembly, and (c) the method of

TABLE V - COMPARATIVE ASSESSMENTS OF ALTERNATIVE METHODS
FOR DRY STORAGE OF SPENT FUEL[a]

	Techno-logical Ranking	Licens-ability Ranking	Cost Assessment	
Unconsolidated Fuel			Ranking	Unit Cost ($/kgU)
Water Pool Storage	4	1	13	345
Cask Storage	3	3	3	117
Drywell Storage				
Canned in Reactor Pool	1	2	5	137
Canned in Separate Facility	2	2	10	257
Silo Storage				
Canned in Reactor Pool	5	4	6	160
Canned in Separate Facility	7	4	9	251
Vault Storage				
Canned in Reactor Pool	2	6	15	419
Canned in Separate Facilty	6	6	16	504
Consolidated Fuel				
Water Pool Storage	10	5	12	301
Cask Storage	11	9	1	110
Drywell Storage				
Canned in Reactor Pool	8	7	2	112
Canned in Separate Facility	12	7	8	240
Silo Storage				
Canned in Reactor Pool	9	8	4	131
Canned in Separate Facility	14	8	7	232
Vault Storage				
Canned in Reactor Pool	13	10	11	300
Canned in Separate Facility	15	10	14	406

[a]Lowest numerical ranking indicates most desirable method, highest numerical ranking represents least desirable method, and so on.

disposition of end fittings and skeletal parts resulting from disassembly activities. Moreover, based on present knowledge, there does not appear to be a sufficient prospective economic advantage for consolidation to risk the technical and licensing uncertainties currently involved.

3.2 Storage in an air-cooled concrete vault (or a water pool) requires a large initial capital investment in facilities which results in very high average unit costs for the storage of spent fuel ($300-500/kgU).

3.3 Storage of spent fuel by modular methods (such as casks, drywells, and silos) where storage capacity need be added only as required, results in the lowest unit costs for storage inasmuch as a minimum initial investment is required; in addition, the risk of installing more capacity than ultimately needed is eliminated.

3.4 If at all possible, canning (and disassembly, in the case of consolidated fuel storage) should be conducted in the reactor storage pool, otherwise storage of uncanned fuel in storage casks is the most attractive storage method. The construction and operation of a separate canning facility requires a significant initial capital investment which results in a large increase in the unit costs for storage (by $85-120/kgU) over that for canning of spent fuel in the reactor pool. This course of action will, however, involve two licensing activities for U. S. facilities, as the canning/disassembly would be licensed under 10CFR50, while the storage facility would be licensed under 10CFR72.

3.5 Storage of spent fuel in drywells or storage in casks appear to represent the most desirable methods for storage at the present time and on balance are approximately equivalent. The cost of storage of unconsolidated fuel in casks was estimated to be $117/kgU, and in drywells was estimated to be $137/kgU, but the uncertainties in capital cost of the casks and in their technical performance characteristics tend to offset this difference.

3.6 At the present time, storage of spent fuel in concrete storage silos does not appear to be as attractive from an economic standpoint as storage in drywells or casks. This is due to the fact that silo storage is in its most economic form when the fuel is canned in the reactor pool -- and this requires equipment for loading the silos with spent fuel, which is more complex and costly than for the other methods.

3.7 Additional research, development, and demonstration work should result in lower costs for storage of spent fuel by modular methods. However, it is our opinion that it is extremely doubtful that such work will be successful in reducing the unit costs below $100/kgU.

4.0 REFERENCES

4.1 Statement by the President [Carter] on [US] Nuclear Power Policy, April 7, 1977.

4.2 O. F. Brown, letter to Marshall E. Miller, Esq., U. S. Nuclear Regulatory Commission, March 27, 1981, Subject: Proposed Rulemaking on the storage and Disposal of Nuclear Waste (Waste Confidence Rulemaking) NRC Docket No. PR-50,51 (44FR61372).

4.3 U.S. Department of Energy, Statement of Position of the United States Department of Energy in the Matter of the Proposed Rulemaking on the Storage and Disposal of Nuclear Waste, DOE/NE-007, April 15, 1980

4.4 P. A. Anderson and H. S. Meyer, Dry Storage of Spent Nuclear Fuel, NUREG/CR-1223, April 1980

4.5 U.S. Energy Research and Development Administration, Waste Management Operations, Idaho National Engineering Laboratory, ERDA-1536, September 1977

4.6　J. D. Hammond, R. S. P'Pool, and R. D. Modrow, Safety Analysis, Peach Bottom Spent Fuel Storage, IN-1465, Idaho National Engineering Laboratory, Idaho Falls, ID

4.7　U.S. Energy Research and Development Administration, Environmental Impact Statement-Receipt, Storage, and Processing of Rover Fuel, WASH 1512, April 1972

4.8　U.S. Nuclear Regulatory Commission, Facilities License Application Record, License Docket 50-267

4.9　U.S. Energy Research and Development Administration, Draft Environmental Statement-HTGR Fuels Reprocessing Facilities, National Reactor Testing Station WASH-1534, January 1974

4.10　J. M. Davis, Demonstration of a Surface Storage System for Spent Fuel or Waste, CONF-771102-18, Atlantic Richfield Hanford Co., Presented at 20th Annual Meeting, American Institute of Chemical Engineers, New York City, NY, November 1977

4.11　K. H. Henry and D. A. Turner, Storage of Spent Unreprocessed Fuel (SURF), CONF-780316-5, Rockwell Hanford Operations, Richland, WA, March 1978

4.12　J. A. Morrison, AECL Experience in Managing Radioactive Wastes from Canadian Nuclear Reactors (AECL-4707, Atomic Energy of Canada, Limited, Chalk River, Ontario (March 1974); Quoted in ERDA-76-43, Alternatives for Managing Wastes from Reactors and Post-Fission Operations in the Nuclear Fuel Cycle, Vol. 3, May 1976

4.13　K. J. Truss, Concrete Canister Demonstration Project, Preliminary Safety Analysis Report, Whiteshell Nuclear Research Establishment, Pinaua, Manitoba, 1975; quoted in ERDA-76-43, Alternatives for Managing Wastes from Reactors and Post Fission Operations in the Nuclear Fuel Cycle Vol,3, 1976

4.14　P. J. Dyne, Managing Nuclear Wastes AECL-5136, Atomic Energy of Canada, Limited, Chalk River, Ontario, May 1975. Quoted in ERDA-76-43, Alternatives for Managing Wastes from Reactors and Post Fission Operations in the Nuclear Fuel Cycle, Vol. 3, 1976

4.15　E. D. Maxwell and D. Deacon, "Dry Storage of Irradiated Magnox Fuel in Air", Nuclear Engineering International, May 1979

4.16　P. L. Gray, Licensing Schedule for Away-From-Reactor (AFR) Spent Fuel Storage Facilities DP-1582 (August 1981), E. I. du Pont de Nemours & Co., Savannah River Laboratory, Aiken, South Carolina

DISCUSSION

M.C. Tanner, United Kingdom

The tabulation of results illustrates dramatically the impact of front end expenditure on unit costs when a high interest rate (14%) is assumed. What would be the effect on the ranking order of unit costs of alternative assumptions about interest rates ? In particular what happens if a "real" interest rate is assumed ?

E.R. Johnson, United States

U.S. utilities are currently paying interest rates on first mortgage bonds in the range 16-18 %. We believe that 14 % is conservatively low for the foreseeable future.

Lower interest rates would reduce the impact of high front end expenditure. To take an extreme, at a 0 % interest rate dry-wells would become more economic than storage casks.

I do not have the answer to specifically what happens if the "real" interest rate is assumed, but the trend cited above (for 0 %) would suggest that the order would not be changed significantly. However, the unit costs developed in our study are subject to escalation in future years at whatever rate of inflation is experienced whereas the use of the so-called "real" interest rate would yield a unit cost that would be subject in its entirety to escalation over the entire period of operation of the facility.

R. Verdant, France

Avez-vous essayé d'étendre vos résultats (les coûts unitaires) au cas d'un stockage centralisé (AFR) de quelques milliers de tonnes d'U pendant quelques dizaines d'années ? Quelle est la durée de stockage AFR prise en compte dans votre étude ?

E.R. Johnson, United States

Our study was directed only at the use of dry storage methods to supplement the capability of the reactor storage basin. We did not consider the larger AFR storage facilities.. and do not believe that similar trends would be experienced.

Our study considered storage commencing in 1985 and continuing through 2009, followed by emptying of the facility in 2010 and 2011. Therefore some fuel would be stored for 25 years (at the most) and some fuel would be stored for 2 years (at the least), with the average being about 13 years.

B.R. Teer, United States

Has anything happened in the months since the study was completed which would change your subjective ratings and/or your conclusions ?

E.R. Johnson, United States

If we were to update the study today we would do the following :
(a) Take into consideration additional R & D on consolidation of spent fuel that has taken place in the last year.

(b) Determine the economic effect of modularizing vault concepts and evaluate different vault concepts.

(c) Review drywell and concrete silo design for possible simplification and appropriately modify the ratings.

(d) Consider current attitudes of U.S. regulatory authorities with respect to maximum fuel clad temperature and stress corrosion cracking and appropriately consider in the ratings.

(e) Consider new data in heat transfer capabilities of storage casks.

C.J. Ospina, Switzerland

Please what is your estimation on extra costing when encapsulating spent fuel for dry storage and what risks can be present during these types of encapsulation operations ?

E.R. Johnson, United States

Encapsulation of unconsolidated fuel is estimated to cost about $ 19/kg U (1981 dollars US) when accomplished in the reactor storage basin.

The risks during encapsulation are negligible in my opinion. There is, however, an operational inconvenience involved in having to encapsulate in the reactor storage basin.

ECONOMICAL DRY STORAGE OF LARGE AMOUNTS OF SPENT FUEL AND VITRIFIED
HIGH LEVEL RADIOACTIVE WASTE

R.F. Bokelmann, R.R. Kühnel, B.J.G. Leidinger
Kraftwerk Union AG
Offenbach, FRG

Abstract

The economical dry storage of spent fuel assemblies and
vitrified high level radioactive waste demands facilities that fulfil, besides safety requirements, especially the following criteria:
o minimization of the specific space required for the storage of
   one fuel assembly or glass cylinder,
o minimization of the number of active systems and components to
   alleviate operation and maintenance,
o application of current materials.
Comprehensive thermodynamical calculations on the single and dual
cycle dry storage facilities, developed by KWU, have shown that the
flow and temperature distributions within the storage rack and the
other parts of the facility determine the feasibility of such dry
storage systems. The parameters are mainly governed by the orientation and the pitch of the storage racks.

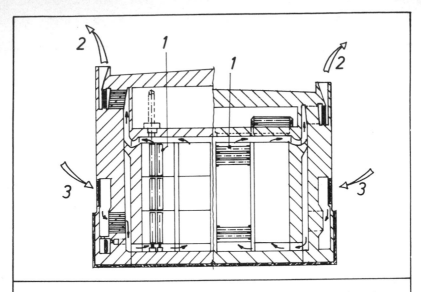

Figure 1 :Dry storage facility for spent fuel
and HLW :
Single cycle mode -
vertical and horizontal orientation

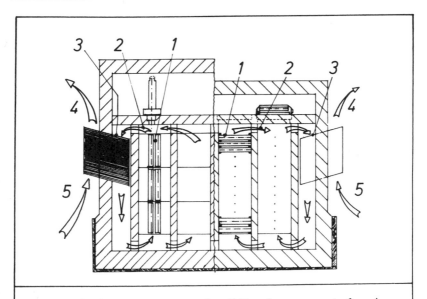

Figure 2 :Dry storage facility for spent fuel
and HLW :
Dual cycle mode -
vertical and horizontal orientation

1. Introduction

KWU has developed inter alia air cooled dry storage facilities in single and dual cycle mode applying inactive cooling systems (e.g. heat pipes) preferably. Specific research and development work has shown that dominant criteria for the engineering and design of such storage facilities are, besides those following the safety requirements, the allowable temperatures of the storage goods and of the internal equipment of the facility as well as radiation exposure of the environment.

This paper features the mentioned two types of dry storage facilities, their specific application and their characteristic advantages.

2. Dry storage facilities developed by KWU

The dry storage concepts described below are applicable for the storage of spent LWR fuel assemblies as well as heat generating glass cylinders containing high level radioactive waste from the reprocessing of spent fuel. The aptitude for the storage of both goods is due to their similarity, e.g. referred to their geometry, their heat generation and radiation level.

2.1 Single cycle dry storage facility

The design is shown in Figure 1, depicting vertical orientation of the storage goods on the left, horizontal orientation on the right side. It fulfils the following essential criteria:
o passive air cooling system, supported by a small fan if necessary,
o sufficient protection against radiation exposure of the environment,
o no disassembling and conditioning of spent fuel assemblies,
o two independent barriers against radioactivity release,
o resistance against external events (e.g. aircraft crash, gas cloud explosion and fire).

The handling of spent fuel assemblies to be stored consists of the following steps:
o unloading the assemblies from the fuel shipping cask,
o packing-in of the fuel assemblies (1 PWR or 2 BWR assemblies per can) including gas-tight welding of the lid and helium leakage test of the can,
o transfer of the can to the loading machine,
o storing of the can in its destinated position,
o closure of the storage shaft by replacing the radiation shielding lid.
(In the case of glass cylinders the canning-in procedure is not performed in the storage facility.)

The first barrier against the radioactive products' release is performed by the fuel pin cladding, which is considered to be intact in the very most cases, when the fuel assemblies are canned in; the second barrier is given by the gas-tight cans.
(In the case of glass cylinders the two barriers are represented by the glass and the gas-tight stainless steel lines which comprise the glass cylinder.)

2.2 Dual cycle dry storage facility

This design is shown in Figure 2, also depicting vertical orientation of the storage goods on the left, horizontal orientation on the right side.
The design criteria are the same as in Chapter 2.1 annotated, with the exception that three instead of two barriers against radioactivity release are existing; the third barrier is performed by the building's wall, which is not penetrated by openings for cooling purposes.

The handling of the storage goods is the same as in the single cycle facility.

The storage area consists of concrete shaft, in which specific supports are built in.

The cooling system consists of an interior gas circuit, which mass flow and temperature distribution is determined by the natural buoyancy in the heated shafts and the pressure drops in the entire circuit. The heat generated in the fuel is mainly transferred to the can-walls by radiation, that in the glass cylinders by heat conduction to the liner.

The heat sink in the internal circuit is performed by the evaporator section of the heat pipes. The heat pipes (which are penetrating the building's wall technically gas-tight) transfer the heat through the wall to the condenser side of the pipes which are cooled by the outdoor coolant flow.

The design of the cooling system is so that after a total loss ob one heat sink (e.g. one of the two heat pipe bundles) the temperature limit of 400 ºC at the hottest spot of all fuel pin claddings or glass cylinders surfaces is not exceeded.

3. Thermodynamical design

The design of dry storage systems is essentially based on thermodynamical aspects referred to the storage goods as well as to the facility and its equipment, because more or less temperature limits are to be considered. By theoretical modelling design criteria and assumptions are to be made to gain reliable results with reasonable effort.

3.1 Design criteria

o To prevent recirculation and bypass effects by partial loading of the storage area the lower connection channel beneath the storage racks has to be dimensioned sufficiently large, whereas the upper connection channel on top of the racks is not allowed to provide additional buoyancy.
o The storage area is completely to be covered with storage racks.
o To gain comparable results, each storage facility is designed with equal storage capacity, guaranteed by equal pile height and width.
o The heighth of a pile is chosen so, that in the case of the dual cycle facility the efficiency of the natural convection flow of the internal coolant circuit is large enough to render heat transfer with significant small temperature differences between can and coolant.
o As a result of this a pile height of either three cans (vertical orientation) or thirty cans (horizontal orientation) one above the other is chosen.
o The storage capacity-determining parameters are the pitch S in the depth direction and the can-diameter D. In the calculations the parameter (S − D) is chosen, because the diameter D has been fixed after optimization following several separate calculations and experiments. In the case of vertical orientation a squared pitch is applied.
o The heat pipe configuration of the dual cycle facility is designed such that each heat pipe transfers the same amount of heat; so the mass flows on the evaporator and condenser sides are equal.

3.2 Assumptions for the mathematical modelling

o One dimensional description of the storage shaft
o Homogeneous storage pattern with utilization of the entire storage space and maximal heat generation
o Homogeneous distribution of the heat sources
o Fully developed flow patterns in all heated and cooled channels

o   All pressure drop and heat transfer effects are described by
    correlations due to forced convection, which can be considered
    as a conservative assumption.

3.3 Discussion of the results

        Following the above-mentioned criteria and assumptions, it
can be stated that the temperatures on several field points as
functions of the storage rack's geometry are essential for design
and operation of a dry storage facility.
Figures 3 to 6 show significant temperatures in dependency of the
variable (S - D) in a range from 0 to 0.4 m. The numerals dedicated
to the several temperature curves correspond with those in Figures 1
and 2 and provide local temperatures in different field points of
the storage facility.
In order to find out, what kind and mode of storage is advantageous
out of the thermodynamical point of view some physical considerations have to be made.

        Decreasing pitch increases the drag of the storage rack and
provides the following countercurrent effects:
i.)   the mass flow of the coolant decreases and the coolant tempera-
      ture increases,
ii.) the turbulence increases and so the heat transfer from the hot
      can to the coolant increases as well.
The influence of these effects to vertical orientation is smaller
than to horizontal orientation, because the drag of vertical orientated racks is smaller than that of horizontal orientated racks.

        Concerning the single cycle and dual cycle storage facility
the consequence of the discussed phenomena can be stated in following manner:
Single cycle storage facility (Figures 3 and 4)
As the drag of the building of the single cycle facility (consisting
of inlet, outlet, filter package and radiation traps) is much larger
than the drag of the storage area, in the case of reasonable pitch
values, the mass flow of the coolant is governed by the building drag
and therefore only the heat transfer effect in the storage rack is
to be considered. As this effect is stronger in horizontal than in
vertical orientation, the storage in horizontal orientation is advantageous in comparison to vertical orientation subject to a single
cycle facility, as far as can surface temperatures are concerned.
Only in the case of technical infeasible pitch-values (S - D < 0,05 m)
the temperatures referred to the horizontal storage increase to unlimited values, because the coolant mass flow approaches to zero.
The monotonous increasing of the can surface temperature against
the coolant temperature with increasing pitch values in the case of
horizontal orientation is due to the fact that turbulence decreases
with increasing pitch.

Dual cycle storage facility (Figures 5 and 6)
The feedback of the drag to the mass flow of the coolant is larger
than in the case of the single cycle system. Therefore, both effects
mentioned above are existing (mass flow reduction and heat transfer
improvement due to drag-increasing). Referred to our design both
effects are nearly equalizing each other and result in the fact
that can surface temperatures are nearly equal due to vertical or
horizontal storage orientation.
As far as coolant temperatures inside the storage building are concerned, the case of vertical orientation provides temperatures some
lower than in the case of horizontal orientation.

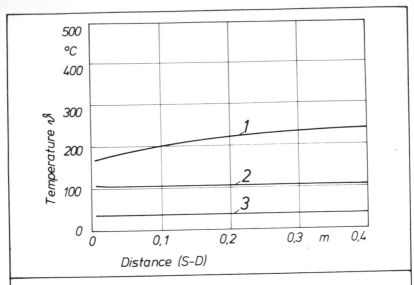

Figure 3 : Single cycle mode - vertical orientation
Fluid and system temperatures
as functions of geometric effects

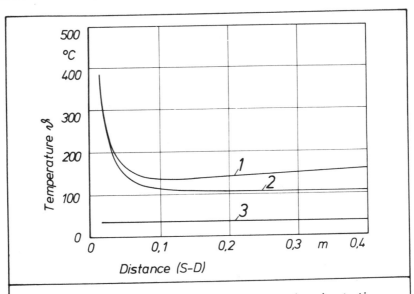

Figure 4 : Single cycle mode - horizontal orientation
Fluid and system temperatures
as functions of geometric effects

4. Economical aspects of the investigated dry storage facilities

It has been pointed out earlier, that the pitch is a parameter which determines the volume of the storage area. The performed calculations have provided the dependency of significant temperatures as functions of the parameter (S - D).
To compare the efficiency of the kind of facilities (single cycle and dual cycle) and of the storage mode (vertical and horizontal orientation of the storage goods) against each other, permitted temperatures (e.g. can surface and coolant temperatures) have to be chosen, followed by the evaluation to corresponding (S - D)-values.

Two important cases are to be considered:
i.) equal can surface temperatures
ii.) technical feasible pitch values
and their influence to the storage building volume.

Case 1: Concerning single cycle facilities equal can surface temperatures due to vertical and horizontal orientation are gained with a (S - D)-value of 0.04 m (see Figures 3 and 4). The storage area volume calculation referred to this pitch including the necessary room for the loading machine renders the result of about 15 per cent in favour towards the vertical versus the horizontal orientation of the storage goods.
This ratio increases, if instead of (S - D) = 0.04 m the pitch is made to zero in the case of vertical orientation, which is theoretical feasible following the temperature curve in Figure 3.
Considering the dual cycle facility, the comparison for a can surface temperature of 270 °C on both the vertical and horizontal orientated can is made. The necessary (S - D) values are 0.24 m and 0.16 m respectively. The storage area volume reduction resulting from this values is about 20 per cent in favour towards the horizontal versus vertical orientation.
The tendency of increasing favour of horizontal versus vertical orientation with increasing pitch is easy understandable, because the vertical orientation of the storage goods is combined with the squared pitch (see Chapter 3.1) and the loading machine area ratio is increasing in the same manner and renders an additional disadvantage of the vertical orientation mode.

Case 2: In order to find a technical feasible pitch value design features have to be respected. To fulfil radiation shielding requirements, each loading opening in the ceiling which devides the storage area from the loading machine area, has to be constructed with a radiation trap (see Figures 1 and 2). So a hole-to-hole pitch of about 0.7 m is necessary, that means a (S - D) of about 0.35 m in the case of vertical orientation of the loading goods. This distance can be reduced by utilizing a staggered loading pattern to a minimum of 0.3 m. Following this value for the vertical orientation and leaving the value of 0.35 m valid for the horizontal orientation, the storage area volume ration is 0.6 in favour of the horizontal orientation versus the vertical orientation, independent of single cycle or dual cycle mode.

Although, as demonstrated especially in the last case, the storage area volumes can differ significantly depending on horizontal or vertical orientation of the storage goods it has to be taken into account, that the contribution of the storage area volume to that of the whole storage facility is only in the range of 30 - 50 per cent. So the influence of the storage area volume increase of a specific amount to the storage facility is to be divided by 2 to 3 to gain the enlargement of the entire building (including FSC-unloading and fuel assembly-handling area).

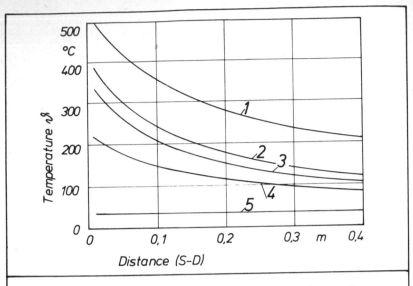

Figure 5: Dual cycle mode - vertical orientation
Fluid and system temperatures
as functions of geometric effects

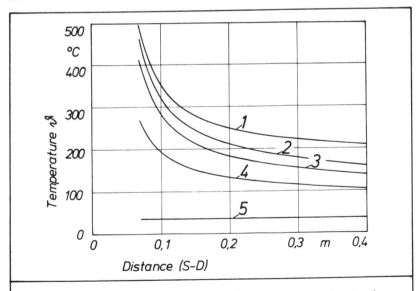

Figure 6: Dual cycle mode - horizontal orientation
Fluid and system temperatures
as functions of geometric effects

5. Conclusions

The performed investigations have demonstrated that the dry storage of spent fuel assemblies and high level vitrified radioactive waste is technically feasible.

Considering the design criteria and assumptions given in Chapter 3, the following can be stated:

o Each storage task can be performed generally with a single or a dual cycle facility in vertical or horizontal orientation of the storage goods (either spent fuel assemblies or glass cylinders);

o No significant economical advantage is due to one kind of facility or storage orientation mode in comparison to the other.

These statements may become invalid if boundary conditions are changed (e.g. the capacity of the facility, the pile dimensions, the allowable surface and coolant temperatures etc.) or if considered safety requirements are omitted.

So the decision in favour of one kind of facility or storage orientation mode can only be made, if further detailed conditions are published.

DISCUSSION

E.O. Maxwell, United Kingdom

In the single cycle facility you showed cooling air inlet and outlet filters. Do you think enough bouyancy head is created to overcome the pressure drop across the filters. In fact I am not sure why you have filters in the system. I assume the fuel assemblies are canistered ?

R.F. Bockelmann, Federal Republic of Germany

The mentioned filters are not absolute filters (e.g. HEPA-quality) but are intended to keep off dust, insects and other particles of significant size.

The operation of the vent, which is foreseen for specific operation conditions, assures the existence of a sufficient coolant flow.

N. Bradley, United Kingdom

You state that conventional materials are used. In the case of concrete, how is this compatible with 130°C air outlet temperature; do you use insulation ?

R.F. Bockelmann, Federal Republic of Germany

It is foreseen to protect those parts of the concrete walls which could be attacked by high temperature coolant (temperatures up to 130°C) by a temperature insulation (e.g. a liner and insulation material).

From our point of view, a wall temperature gradient of 30°C/m is acceptable utilizing heavy concrete as used in nuclear constructions.

A.B. Johnson, United States

You indicated carbon steel as a principal material and advised against consideration of expensive materials. Do you regard stainless steel as too expensive for any use in the system ?

R.F. Bockelmann, Federal Republic of Germany

Yes !

LE STOCKAGE A SEC DES COMBUSTIBLES IRRADIES

EXPERIENCE FRANCAISE

Auteurs :

A.S. Annie SUGIER
Commissariat à l'Energie Atomique (DgAEN)
75015 PARIS

J.G. Jean GEFFROY
Commissariat à l'Energie Atomique (DEMT)
CEN-SACLAY

M.D. Michel DOBREMELLE
Commissarait à l'Energie Atomique (DEMT)
CEN-SACLAY

Avec la collaboration de MM. LANGLOIS (DIP), et VERDANT (DPg).

L'option choisie par le France pour la gestion de ses combustibles irradiés est le retraitement, après un stockage préalable de courte durée en piscine.

Le groupe CEA n'a donc pas développé pour son programme de stockage intérimaire des combustibles de réacteurs à eau type PWR, le stockage à sec. Il a cependant acquis une expérience certaine dans ce domaine tant au stade des études en amont de la réalisation dans ses laboratoires d'études thermiques et mécaniques, qu'au stade de la conception, de la mise au point, de la réalisation et de l'exploitation de telles installations ou d'ouvrages semblables pour des objectifs voisins. La présente communication a pour objectif de décrire cette expérience et de proposer une base de réflexion pour une étude de préfaisabilité d'une installation de stockage à sec.

I. Remarques préliminaires : le choix français.

La France a choisi de retraiter ses combustibles irradiés et de proposer ses services à des clients étrangers. Le retraitement a pour objectif une gestion rationnelle des combustibles irradiés. Il permet en effet de séparer les matières recyclables dans des processus de production d'énergie des déchets qui seront conditionnés et stockés selon leur niveau d'activité.

Le Gouvernement et le Parlement, au cours du débat énergétique national de l'automne 1981, ont décidé la poursuite de l'effort nucléaire et ont autorisé l'extension de la capacité de l'usine de retraitement de La Hague jusqu'à 1600 t/an pour satisfaire à la fois les besoins nationaux rapidement croissants et nos engagements internationaux. Il faut rappeler que la France est le 2ème pays producteur d'énergie nucléaire dans le monde derrière les Etats-Unis.

Le retraitement des combustibles irradiés implique un stockage limité dans le temps en amont des usines (stockage de refroidissement et stockage tampon). Le stockage sous eau a été retenu en raison de la maîtrise du procédé, de la sûreté potentielle que représente le volume d'eau des piscines du point de vue de la protection biologique et du point de vue thermique ainsi que de l'intérêt, pour les controles internationaux de pouvoir surveiller visuellement les assemblages stockés.

Dès à présent, à La Hague, la réception des combustibles est engagée dans les nouvelles piscines de stockage intermédiaire, d'une capacité unitaire de 2000 tonnes. Signalons, en effet, la livraison à La Hague en avril 1982 de la 1000ème tonne de combustibles irradiés en provenance de réacteurs européens à eau ordinaire.Une autre piscine d'une capacité totale identique est en construction (mise en service en 1983), deux installations d'une capacité de 2000 t chacune sont à l'état de projet. Le coût de la première tranche actuellement en service (y compris l'installation de réception des combustibles) est de l'ordre de 800 MF courants pour une installation mise en service en 1982.

Ainsi la France contribue à limiter les tonnages de combustibles irradiés devant être placés dans des stockages de longue durée.

S'agissant des pays de la CCE, notre rôle apparaît clairement dans les deux tableaux suivants extraits d'un rapport de la CCE publié en février 1982 et réactualisé par le CEA pour ce qui concerne les informations relatives à la France (rapport COM 82/37 final - 5/2/82- Comité consultatif ad hoc en matière de retraitement des combustibles nucléaires irradiés - Communication de la commission du Conseil).

1°) Capacités de retraitement cumulées disponibles (usines mises en service, engagées et prévues) en tonnes d'uranium.

	1985	1990	1995	2000
France	1.400	4.400	12.000	20.000
Total CCE	1.470	6.885	21.435	39.085

2°) Quantités de combustibles stockés non couvertes par les capacités de retraitement disponibles aux dates indiquées précédemment: [1]

1985	1990	1995	2000
8.000	16.000	18.000	24.000

La mise en service progressive entre 1987 et 1989 des extensions UP3 et UP2 800 des usines de La Hague, contribuera pour près de 60% à la capacité de retraitement disponible dans la CCE en 1995 ce qui limitera à 18.000 t les tonnages de combustibles irradiés cumulés non couverts par le retraitement à cette date et à 24.000 t en l'an 2000. On conçoit, cependant, qu'il soit judicieux de s'interroger sur le type d'extension des capacités de stockage existantes permettant de satisfaire ces besoins excédentaires en installations de stockage. Ce problème ne se pose pas pour la France compte tenu de ses capacités nationales de retraitement et de stockage tampon.

(1) Compte tenu des combustibles acheminés vers la CCE pour y être retraités.

II. L'expérience acquise par le groupe CEA dans le domaine du stockage à sec.

Bien que le groupe CEA n'ait pas développé le stockage à sec pour son programme de stockage intérimaire des combustibles de réacteurs à eau type PWR, il a acquis une expérience certaine dans ce domaine tant au stade des études en amont de la réalisation dans ses laboratoires d'études thermiques et mécaniques, qu'au stade de la conception, de la mise au point, de la réalisation et de l'exploitation de telles installations ou d'ouvrages semblables pour des objectifs voisins :

- châteaux pour le transport à sec des combustibles PWR,
- installation de stockage à sec des combustibles irradiés des réacteurs surgénérateurs en amont du pilote de retraitement TOR (Traitement Oxyde Rapide),
- installation de stockage à sec des déchets vitrifiés de haute activité à Marcoule AVM et à La Hague (AVH en projet).

Avant d'analyser quelles pourraient être les options majeures d'un projet plus spécifique de stockage à sec des combustibles irradiés des réacteurs à eau ordinaire s'appuyant sur l'expérience acquise par le groupe CEA, nous présenterons tout d'abord les moyens d'essais développés au CEA et les techniques mises au point.

Moyens de recherche et développpenet.

Différentes options de stockage à sec sont envisageables. Parmi les principales, nous citerons, d'une part la mise en structure refroidies par ventilation forcée du combustible nu et, d'autre part, le refroidissement sans machine du combustible mis en conteneur. Dans le cas du soufflage, le groupe CEA s'appuie sur les études et moyens qui ont été développés pour la filière française graphite Gaz. Dans le deuxième cas le groupe CEA se réfère aux moyens mis en oeuvre pour la conception des châteaux de transport.

Les grandes fonctions de l'emballage, qu'il soit destiné au transport ou au stockage, sont de même nature. Il s'agit de confiner, d'assurer une protection biologique, d'éviter le risque de criticité, de refroidir, et de permettre la manutention. Des problèmes plus spécifiques peuvent cependant se poser dans le cas d'une immobilisation de très longue durée.

Le groupe CEA dispose, pour définir ces différentes fonctions d'un potentiel important d'étude, de calcul et d'expérimentation.

Nous n'insisterons pas sur les codes et méthodes de calcul mis à la disposition du concepteur. Un apport indispensable qu'il nous semble important de souligner est celui des possibilités d'expérimentation. En effet, les configurations géométriques et les phénomènes d'échange étant complexes, il faut avoir recours à des expériences pour valider les procédés envisagés. Des maquettes chauffantes ont été réalisées pour simuler les combustibles en conteneur ou en emballage. La figure 1 montre une maquette chauffante de combustible type eau ordinaire qui peut se monter dans une cavité représentant une portion d'emballage. Dans cet exemple, le dégagement de chaleur est obtenu à l'aide de résistances internes aux aiguilles. Ces études menées pour différentes natures et pressions de gaz aboutissent à la validation des modèles d'échanges utilisés dans les codes de calcul. Les principaux paramètres de l'étude sont le nombre des aiguilles par étui, le diamètre de l'étui, la puissance dissipée, l'arrangement des aiguilles dans l'étui les positions verticales ou horizontales de l'étui, les conditions aux limites thermiques. Des bancs d'essais sont également réalisés permettant d'étudier l'effet des ailettes externes pour refroidissement externe.

Enfin, il faut signaler que des expériences directes sur des aiguilles irradiées équipées de thermocouples dans des laboratoires chauds permettent d'obtenir de précieux résultats.

Expérience acquise par le groupe C.E.A.

1) Transport à sec des combustibles irradiés LWR. Le groupe CEA a participé à la mise au point des châteaux et à l'exécution avec ses partenaires de transports de combustibles irradiés (Nuclear Transport Limited pour les transports internationaux en Europe, Pacific Nuclear Transport Limited pour les transports par mer entre le Japon et l'Europe).

Figure 1. Maquette chauffante "PWR"

L'expérience acquise depuis plus de 15 ans en France avec les châteaux à sec pour le transport des combustibles oxyde et métal est très satisfaisante. Ce choix s'explique pour des raisons :
- de sécurité : à 200°C, la pression de l'eau est de 15 bars et les gaz de radiolyse peuvent encore l'augmenter. Pression et gaz de radiolyse compliquent le déchargement et vont à l'encontre de la sécurité
- d'économie : les opérations de décontamination au déchargement sont simplifiées d'où économie de temps (rotations des châteaux) et de personnel (hommes-rem).

Les châteaux standards qui servent au transport à sec des combustibles LWR sont cylindriques,ils pèsent 100/110 tonnes, le panier intérieur permet de recevoir douze éléments combustibles type PWR répartis en quatre secteurs (cf fig.2).

Les emballages comportent une protection gamma, une protection neutronique, une protection thermique, un dispositif pour éviter les accidents de criticité et un dispositif de refroidissement pouvant évacuer jusqu'à 100 Kw (chaque assemblage transporté a une puissance résiduelle de 6 à 12 Kw). La protection gamma est assurée par l'utilisation d'acier forgé, la protection neutronique par des matériaux légers (corps hydrogénés) sur l'extérieur. Le combustible est lui-même placé dans une première enveloppe métallique cylindrique étanche à fond fixe. La fonction de refroidissement doit permettre l'évacuation de la puissance résiduelle avec la double contrainte de base suivante : respect de la température maximale admissible à la surface extérieure du colis (82°C règlementaires) et respect de la température maximale admissible pour le combustible (550°C au point le plus chaud de la gaine soit 12 Kw/ass; en fait, un maximum raisonnable a été fixé 450°C). Le refroidissement ne s'effectue que de façon intrinsèque, c'est-à-dire sans machine comme un ventilateur ou une pompe, le château étant placé à l'horizontale. On ne rencontre donc que trois modes de transfert thermique : la conduction dans la structure, le rayonnement entre les structures dans les cavités et vers l'environnement, la convection naturelle en milieu confiné et vers l'environnement.

Le prix actuel d'un tel emballage de transport est de l'ordre de 7 MF. En admettant qu'un emballage de stockage intermédiaire soit plus simple qu'un emballage de transport et qu'un effet de série soit obtenu, le prix d'un emballage de stockage de grande capacité pourrait descendre à 5 MF. Soit environ 900 F/kg d'u pour l'emballage seul.

Il conviendra, en outre, de tenir compte des exigences suivantes imposées par le retraiteur (et/ou le transporteur) pour la reprise ultérieure des combustibles irradiés ou le conditionnement définitif en stockage géologique :
- mise en place de l'unité logistique de stockage des emballages,
- nécessité de prévoir sur le site de stockage une cellule d'examen du combustible avant son acheminement vers sa destination finale afin de connaître sa tenue au stockage.
- nécessité de prévoir et d'étudier une unité de traitement des déchets gazeux et solides, résultant d'un stockage prolongé dans ce type de stockage.

2) Stockage à sec des combustibles des réacteurs surgénérateurs

Le développement des réacteurs surgénérateurs a conduit le groupe CEA à disposer d'une expérience dans le domaine du retraitement et du stockage de ce type de combustible.

L'installation T.O.R. (traitement oxyde rapide) conçue comme une extension de l'atelier pilote de Marcoule (SAP) permettra de porter la capacité de cette installation de 2,3 t/an à 5 t/an. Elle servira également d'unité de recherche et de développement, et comportera une installation amont de stockage des combustibles irradiés.

La mise en service de l'ensemble est prévue en 1985. Le CEA est maître d'ouvrage, la maîtrise d'oeuvre est asurée par une équipe "intégrée" formée de personnel du CEA et de personnel de la Société Générale pour les Techniques Nouvelles (SGN).

Le choix du stockage à sec a été retenu en raison de la capacité modeste du stockage et du souhait de ne pas avoir d'immersion de château. En effet cette opération conduit à mettre en place des installations de traitement d'eau disproportionnées par rapport à la taille du stockage.

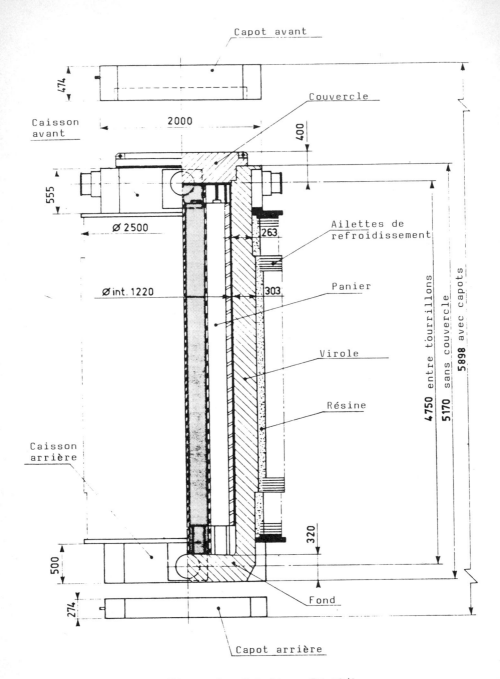

Figure 2 : Emballage TN 12/1

La figure 3 présente une vue schématique du stockage TOR. La capacité totale du stockage est de 9.525 Kg d'oxyde. Le stockage des combustibles se fait dans des puits verticaux ventilés de bas en haut (77 puits en 7 files de 11 puits). Les combustibles sont placés dans des étuis étanches. Chaque puits peut contenir 5 étuis Phénix ou 3 étuis Super Phénix. Un étui contient un ensemble de 93 aiguilles. Les puits ont un diamètre de Ø 300 mm, une hauteur de 9,85 m. Le dégagement thermique par puits varie de 4 à 1 Kw suivant le temps de refroidissement préalable du combustible (6 à 24 mois). La température maximale de gaine est de 640°C.

La ventilation est forcée (1 ventilateur en service, 1 de secours), la variation de débit est réalisée en fonction du chargement par mise en place d'un orifice à chaque puits. Le débit total d'air est de moins de 60.000 m3/h. L'épaisseur de béton nécessaire entre chaque rangée de puits a été définie de telle sorte que la sous-criticité du stockage soit garantie.

Les contraintes limites de température sont, en fonctionnement normal de 33°C pour l'air à l'entrée, 60°C pour le béton et 145°C pour l'étui. En régime accidentel (ventilation de secours) la température du béton peut monter à 80°C. L'air sort entre 50 et 60°C suivant le régime.

La dimension totale de l'installation est une section horizontale de 3,9 m x 4,5 m x 11 m de hauteur et en salle de manutention au-dessus de 3,9 m x 14 m x 5,62m de hauteur.

Les moyens de manutention sont constitués d'un pont de transfert, 2 hublots, 1 poste maître esclave, 1 sas d'entrée des étuis.

Le coût de l'installation seule est d'environ 11 MF. Son adaptation pour le stockage de combustibles type PWR impliquerait des adjonctions et modifications importantes qui ne permettent pas d'extrapoler le coût sans étude approfondie.

3) Stockage des déchets de haute activité vitrifiés

Les déchets vitrifiés de haute activité doivent être stockés pour être refroidis pendant plusieurs dizaines d'années avant leur enfouissement en couches géologiques. La solution retenue pour ce stockage transitoire est le stockage dans des structures de béton refroidis à l'air par ventilation forcée (cf fig.4).

a) Cas AVM

Une première installation de ce type est en service depuis 1978 à Marcoule dans le bâtiment AVM (Atelier Vitrification Marcoule). Ce stockage est conçu de telle sorte que les coûts d'exploitation soient faibles et que l'installation présente toute sécurité tout en permettant la possibilité d'une reprise du verre. Les conteneurs de stockage du verre sont en acier inoxydable étanches et exempts de contamination extérieure. On utilise un refroidissement à l'air qui élimine tout problème de corrosion. On dispose de la possibilité de substituer la convection naturelle à la convection forcée au bout d'un temps de stockage suffisant.

Les conteneurs sont empilés en colonnes dans une fosse, à l'intérieur de "puits" métalliques, sortes de tuyaux suspendus par le haut, et ne reposant pas sur le socle, de 10 m de hauteur et 60 cm de diamètre. Ces parois métalliques jouent un rôle de protection thermique de la structure bétonnée de la fosse. L'air pénètre dans la fosse, passe entre les puits métalliques, descend vers le fond et remonte à l'intérieur des puits le long des conteneurs qu'il refroidit pour ressortir du côté opposé de son entrée. Aspiré par des ventilateurs, il traverse des filtres avant d'être rejeté par les cheminées. L'émission calorifique de chaque puits est de 13Kw. L'ouvrage enterré est à structure de béton armé, d'une longueur totale de 28,5 m sur une largeur de 20,2 m pour une hauteur de 15,2 m.

La capacité de l'installation est de 2.200 conteneurs (soit 330 m3 de verre) répartis en trois tranches.

Deux contraintes limitent la température : la tenue du béton de structure (100°C), la température du verre (500°C à coeur). L'air sort en fonctionnement normal à 110°C et en cas de tirage naturel à 140°C.

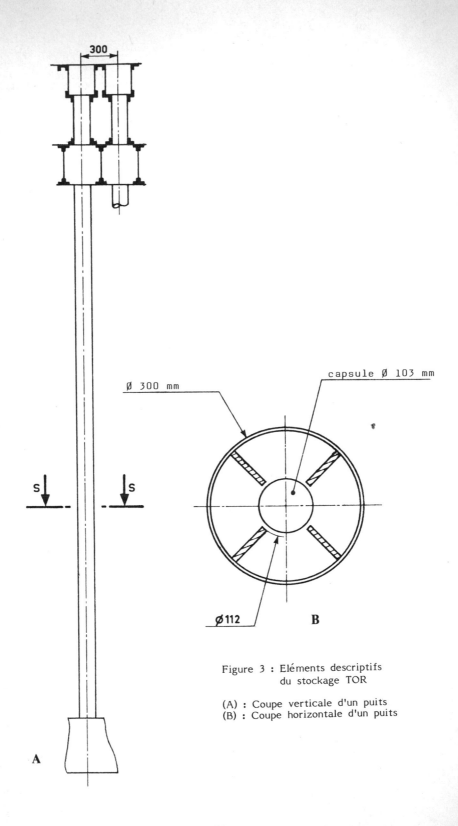

300

Ø 300 mm

capsule Ø 103 mm

Ø 112

B

A

Figure 3 : Eléments descriptifs
du stockage TOR

(A) : Coupe verticale d'un puits
(B) : Coupe horizontale d'un puits

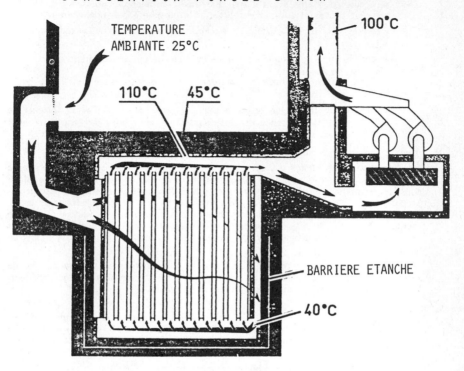

Figure 4 : Stockage type AVM

Pour leur manutention, les récipients de stockage sont transportés sous château de plomb à l'aide d'un portique qui dessert, à partir de la cellule de sortie de l'atelier de vitrification, l'ensemble des tranches du stockage et qui permet de plus, la manutention des bouchons de puits et du sas d'introduction.

Les tâches d'exploitation concernent la manutention des récipients de stockage, le réglage des débits d'air de réfrigération et le changement éventuel des filtres d'entrée et de sortie. En outre chaque puits est équipé d'une mesure de température qui doit être surveillée.

b) Cas AVH

Sur le site de La Hague, les solutions étudiées pour l'installation de stockage des déchets vitrifiés des trois chaines de vitrification des usines UP2 800 et UP3 sont semblables à celles de AVM.

La capacité totale prévue est de 4500 conteneurs répartis en 5 tranches comportant chacune 100 puits. Les châteaux de transfert et l'équipement de manutention sont installés dans la partie supérieure des bâtiments.

Le rôle de l'unité est de recevoir les conteneurs provenant de la vitrification des produits de fission issus de trois chaines de production, d'entreposer ces conteneurs pendant quelques années en assurant leur refroidissement et de permettre leur reprise pour les transférer vers un stockage définitif.

Les conteneurs sont empilés dans des puits constitués par des tubes métalliques verticaux d'un diamètre tel qu'ils ménagent un espace annulaire entre leur paroi interne et celle du conteneur.

La hauteur des puits est de 11 mètres et chaque puits contient 9 conteneurs. L'énergie thermique dégagée dans les conteneurs par les produits de fission est évacuée par une circulation ascendante d'air de refroidissement dans l'espace annulaire ménagé entre conteneurs et paroi du puits.

L'air de refroidissement est prélevé dans l'atmosphère, il refroidit les structures de béton avant d'être distribué à la base des puits. Il est collecté en haut des puits pour être rejeté à l'atmosphère par une cheminée.

Comme dans le cas AVM, les paramètres principaux de l'étude sont les suivants :

- Température maximale de sortie de l'air de refroidissement en tirage forcé =   110°C
- Température maximale de sortie de l'air de refroidissement par tirage naturel =   140°C

La puissance thermique dégagée par puits est de 31,5 Kw. Le circuit de refroidissement est conçu de telle sorte que les modes de fonctionnement soient possibles avec une seule cheminée commune aux cinq modules.

III. Analyse des options envisageables pour le stockage à sec de combustibles PWR.

La plupart des études qui ont été présentées sur le sujet se sont attachées à résoudre les problèmes posés par l'évacuation des calories d'un assemblage placé dans un conteneur étanche rempli de gaz. L'examen critique des solutions proposées montre que l'utilisation de cette technique pour un stockage inter-médiaire conduit à augmenter considérablement la capacité de stockage en piscine réacteur. En effet de tels conteneurs ne peuvent être refroidis aisément que s'ils contiennent des assemblages stockés préalablement pendant plus de deux ans en piscine.

Une alternative à cette solution consiste à augmenter le flux thermique de refroidissement en coulant un métal conducteur entre l'assemblage et le conteneur. Mais cette option a l'inconvénient d'être irréversible.

Si l'on veut véritablement considérer un stockage à sec comme solution d'attente de longue durée avant décision de retraitement, il faut examiner plus particulièrement les possibilités de stockage d'un élément combustible irradié nu. Les conditions d'une telle solution seraient les suivantes =

- assemblages stockés 9 mois en piscine réacteur (7Kw),
- prise en compte des seuls éléments non ruptés. Les éléments ruptés sont conditionnés en bouteilles étanches pour envoi en stockage piscine spécialisée (1%),
- utilisation systématique de la ventilation forcée en marche normale, secours avec filtres très haute efficacité (THE), et adoption d'un circuit à faible perte de charge (température de l'élément combustible limitée à 75°c),
- adoption d'un système de ventilation de secours d'une fiabilité compatible avec son importance pour la sûreté,
- température de l'élément combustible limitée à 125°c dans le cas du fonctionnement du système de secours,
- limitation à 300°c de la température de l'élément combustible en cas de panne du système de secours. Le circuit de tirage naturel circulant sur un réseau de filtres "ad hoc" propres en réserve,
- capacité nominale de 2000T d'U,
- rythme de chargement de 1000T/an,
- durée du stockage 30 ans.

Une telle installation comprendrait une aire de stockage déchargement, une unité de déchargement à sec semblable à celle actuellement étudiée dans le cadre d'une des piscines citées en introduction, une unité de lavage et observations des assemblages entrée et sortie, une unité de stockage en piscine de transfert des éléments ruptés, une unité de traitement de gaz et boues de lavage, une unité de stockage en puits métalliques semblables à ceux adoptés pour AVH à raison de deux éléments par puits, une unité de ventilation forcée avec filtres THE et une cheminée de 60 M.

Une première approximation de l'investissement nécessaire indique une dépense de l'ordre de 1.200 F/kg U ± 40%.

CONCLUSION

L'intérêt de cette approche est de prendre en compte l'ensemble des difficultés et de ne pas repousser à une étape préalable ou ultérieure le traitement des questions inhérentes à ce type de problèmes.

Le groupe CEA dispose d'expériences complémentaires et variées qui lui donnent la possibilité d'aborder des études de préfaisabilité sur un cas précis afin d'apprécier l'intérêt économique comparé des options techniques de stockage à sec généralement envisagées.

DISCUSSION

A.B. Johnson, United States

What is the longest residence for LWR fuel in a dry cask during shipment ?

A. Sugier, France

Environ 6 mois.

A.B. Johnson, United States

When will storage in TOR begin ?

A. Sugier, France

In 1983.

M.S.T. Price, United Kingdom

In your presentation you mentioned that for the TN 12 the peak clad temperature for PWR fuel could be over 500°C. Could you give the technical basis for such a temperature. It would be instructive to give the time that a temperature of say, 300°C, 400°C, 450°C and 500°C could be accepted.

A. Sugier, France

La température limite de 540°C pour des durées de transport de l'ordre du mois a été fixée sur la base d'études métallurgiques afin d'éviter les éventuelles ruptures de gaine au terme du transport au cours du déchargement. La Cogema a cependant préféré ne pas dépasser la température de 450°C. Le temps de stockage préalable en piscine de réacteur est défini en conséquence. Le temps de séjour le plus long en château est d'environ 3 mois.

Confirmant les résultats des spécialistes français en métallurgie je vous signale une publication datant d'avril 1982 de Nuclear Technology faisant état d'essais post irradiation sur les mécanismes de rupture de gaine menés sur des barreaux PWR dont la température était de 482, 510 et 571°C pendant 1 an. Aucune fissure n'a été observée.

THE TN 1300 SHIPPING/STORAGE CASK

SYSTEM FOR SPENT FUEL

R. Christ
W. Anspach
Transnuklear GmbH
D-6450 Hanau 11

ABSTRACT

Based upon its experience gained by development and operation ot its transport casks, Transnuklear has designed a series of casks expecially dedicated to dry intermediate storage. In the paper the design features, manufacturing of prototypes and the status of testing and licensing will be described. It is shown that the system is applicable for storage of different types of fuel.

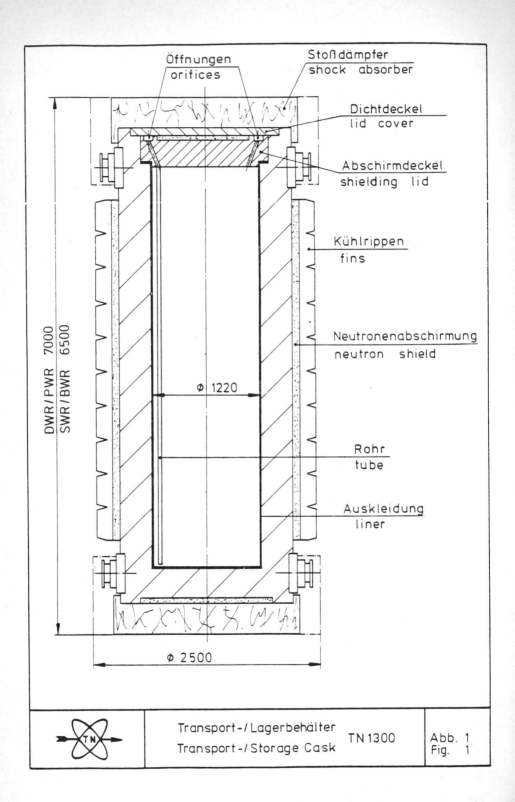

Transport-/Lagerbehälter
Transport-/Storage Cask

TN 1300

Abb. 1
Fig. 1

1. Introduction

Of the various ad-reactor (AR) and away-from-reactor (AFR)
storage techniques which have been studied the intermediate
storage in type B casks is an interesting and viable solution to the problem.

The casks in such a storage system have equally transport
and storage functions. Therefore, they obviously must fulfil
the IAEA-regulations as type B transport casks. In addition,
however, they have to ensure safe storage of spent fuel and
to maintain their integrity under long-term conditions. This
requires inherently safe casks without fragile auxillary
systems, but requires also their adaptation to specific storage
conditions.

For almost 15 years Transnuklear and its foreign partner companies have strictly followed the line of inherently safe
"dry" transport casks filled with air or inert gas and operating
with passive cooling systems, i.e. heat dissipation by natural
air convection. These casks have been successfully operated
for the shipment of spent fuel from European nuclear power
plants. The new generation of forged steel casks type TN 12
and TN 17, about 40 of which are actually in operation, in
production or on order, are used for shipments of European
and Japanese fuel elements to the COGEMA reprocessing plant
at La Hague.

2. Design features of the TN 1300 cask

Based upon the experience gained by development and operation
of its casks, Transnuklear has designed a series of casks
especially dedicated to medium and long-term storage of various
types of spent fuel. The dimensions of these casks were chosen
in such a way as to ensure maximum capacity having in mind the
load and size limitations of specific power plants.

The largest of this group of casks, the TN 1300, is designed
for transport and storage of spent fuel elements from the
1300 MW class of light water reactors. It has a maximum capacity of 12 PWR and 33 BWR fuel elements, respectively. The
design heat load is 50 kW.

Principle dimensions of the packaging are given in figure 1.
The weight of the loaden cask is about 122 t in transport conditions and 115 t in storage conditions.

The cask is built as follows:

- The cask body is made of ductile cast iron (German standard
  GGG 40.3) with cast fins. The body has both containment and
  γ-shielding functions.

- The cylindrical cavity is completely covered by a stainless
  steel liner, 8 mm thick.

- The cask comprises a double lid system with a shielding lid
  (inside) and a sealing lid (outside). Both lids are equipped
  with metal double O-ring sealings with extremely low leak
  rates of about 10^{-8} mbar l/sec. These metal sealings remain
  fully functional even under impact conditions.

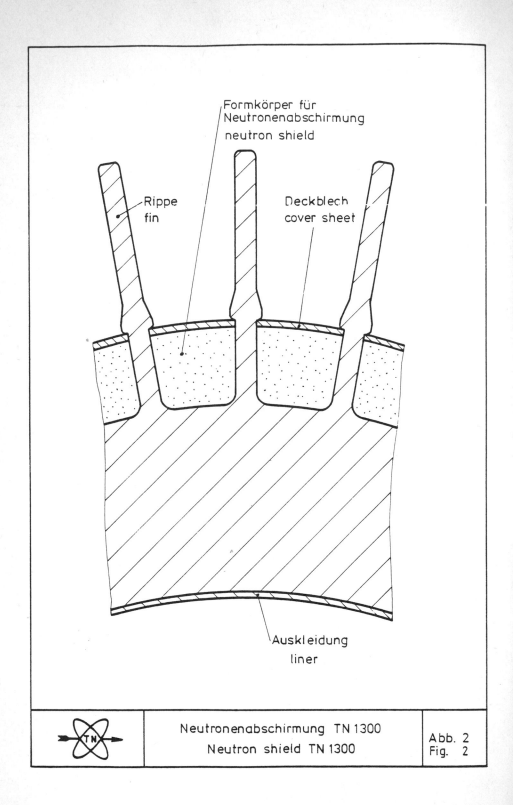

Formkörper für
Neutronenabschirmung
neutron shield

Rippe
fin

Deckblech
cover sheet

Auskleidung
liner

| | Neutronenabschirmung TN 1300 | Abb. 2 |
| | Neutron shield TN 1300 | Fig. 2 |

This high quality sealing system together with an under-pressure of about 0.5 bar in the cavity ensures safe and tight containment of the radioactive content over an extended period of time. Activity release is practically zero.

- The orifice connections through which the container is filled or emptied, pass through the shielding lid and are protected and sealed independently. One of these tubes leads through the basket to the lowest point of the cask. This design avoids any passage through the cask body and thus contributes to safety and low cost production.

- The neutron shielding is located at the outside of the cask between the fins. It is fully encapsulated (fig. 2). The container is so designed that the total dose rate (gamma and neutrons) at the surface does not exceed 20 mrem/h, with the neutron dose rate less than 10 mrem/h.

- The removable basket consists of an aluminium/stainless steel structure. Sintered plates containing boron in the form of B_4C serve as neutron poison for criticality control.

  The choice of different baskets always allows storage of fuel at a maximum fuel rod temperature of less than 400 °C, a value which is generally considered tolerable.

- There are several options for checking tightness during storage without affecting the containment itself. One method requested by DWK for their Gorleben storage facility is to pressurize the interspace between the two lids and to record this pressure remotely.

3. Manufacturing

Three castings of storage casks have been made up to now. The first TN 1300 was cast May 1980 at Thyssen-Gießerei AG in Mülheim/Ruhr. The cask body was removed from the mould after three weeks and was ready for quality control after another two weeks. The quality control tests which were performed in close co-operation with licensing authorities gave very good results for mechanical properties and integrity of the casting of body and fins. For further qualification of the material and the casting method a second TN 1300 was cast in October 1981. Samples for destructive material testing were taken from different sections of the thick walled casting. These tests again demonstrated the material properties to be adequate for the intended use.
A third 20 t casting was produced in February 1982 for a storage cask which will be used for experimental storage of pebble bed reactor fuel (AVR).

4. Testing and licensing

Two 1:3 cask models were produced for the type B tests. In addition to normal drop tests the casks were drop-tested at minus 40 °C and with an artificially damaged wall, namely with a cut of 1/3 of the wall thickness. No cracks and no loss of tightness were observed.

Fire tests were also performed. These tests confirmed the low response of the cask to the fire environment as predicted by computer code and showed the neutron shield to be amost un-affected by the fire.

The complete transport safety analysis report for the TN 1300 version was submitted February 1981. Type B license by German authorities is expected for May 1982.

Additional calculations and tests for demonstration of comp-liance with storage requirements are presently under way. Among this the performance of a full size TN 1300 in a simu-lated air plane crash will be tested in June 1982.

5. Application of the basic design to different types of fuel

The TN 1300 cask is designed for the PWR and BWR fuel from power stations of the 1100 - 1300 MW class. For smaller re-actors with relevant load and size restrictions the TN 900 cask with a capacity of 7 PWR or 17 BWR assemblies is appli-cable.

Cask for mixed oxide fuel are also developped. They mainly differ in the thickness of the neutron shield which must be increased over that for uranium fuel.

An important aspect for the application of the basic design is the cooling time prior to storage. The present situation in Germany is such that short cooling periods have to be taken into account. Therefore external cooling fins and baskets with high heat transfer capabilities are necessary. On the other hand there are countries where due to extended at-reactor storage the cooling period will be more than five years, for example in the United States. For this application we have designed storage casks with a capacity of up to 21 PWR assemb-lies, having a plane external surface and a simplified borated stainless steel basket.

Another promising application is the storage of fuel elements from the German pebble bed reactor. No fins and no neutron shield is necessary in this case. The 20 t prototype cask will be delivered shortly for experimental storage with active fuel at the Jülich research centre.

To complete the picture, evaluations have also been made for fast breeder fuel assemblies. It could be shown that cask storage is feasable also for this type of fuel.

DISCUSSION

H.J. Wervers, The Netherlands

With respect to TN 1300 cask : what is the nature of the neutron shield material ?

R. Christ, Federal Republic of Germany

It is in principle a resin with the addition of material with a high hydrogen content. It also contains 1 % boron.

S.J. Naqvi, Canada

Have these casks been tested to meet the -40°C temperature criterion for impact resistance ?

R. Christ, Federal Republic of Germany

Yes. Evaluation of the low temperature performance is part of the licensing procedure.

S.J. Naqvi, Canada

For how long could fuel be stored in these casks ?

R. Christ, Federal Republic of Germany

At least several decades.

R. Herranz, Spain

Were the TN cask drop tests performed with a full size cask or with a scale model ? What are the requirements of the German authorities in this matter ?

R. Christ, Federal Republic of Germany

Drop tests were performed with a 1/3 scale model. In addition stress calculations were performed to justify the performance of a full size cask in drop conditions. This combination of tests and calculations was accepted by the licensing authorities.

R. Verdant, France

If your casks were to be used for sea transportation, how long could they withstand seawater without leakage problems in the case of an accident at sea (sinking of the ship).

I have the same question about other casks, planned for use in sea transportation, where those mentioned above are not designed for sea transport.

R. Christ, Federal Republic of Germany

We have not studied this problem for the storage casks. For forged steel transport casks, a preliminary assessment has been made which showed that under very conservative assumptions the cask would remain gas-tight for several years and that for several decades there would be no major exposure of the assemblies to seawater.

J. Haddon, United Kingdom

It seems that the TN 12 Flask has been developed fully, is licensable and is ready for use. Can you indicate the cost ?

R. Christ, Federal Republic of Germany

I refer to the presentation of Annie Sugier, where a figure of 7 million FF was given.

H.J. Wervers, The Netherlands

What will be the commercial attitude of Transnuclear with respect to prospective customers ? Will containers be on sale or leased ? Or will Transnuclear deal with the complete "Entsorgung" ?

R. Christ, Federal Republic of Germany

Normally the casks will be on sale, but we are flexible; other arrangements are also possible.

A.B. Johnson, United States

Have aluminium racks been fully qualified for safety aspects when used in casks ?

R. Christ, Federal Republic of Germany

Yes, the TN aluminium baskets are designed to withstand the mechanical and thermal loads of the type B tests in order to maintain criticality control and heat transfer functions.

GENERAL DISCUSSION   DISCUSSION GENERALE

<u>P.G.K. Doroszlai</u>, Switzerland

Our studies arrive at about the same specific storage costs
for storage in casks. However, we found it more economic to take the
canister out of the cask and store it in a multiple concrete silo with
indirect cooling (MODREX). We do not understand why you arrive at
much higher specific costs for silo storage. Did you ever consider
a MODREX type storage with the option of canister transfer between
casks and silo modules, and re-use of casks at other facilities ?

<u>E.R. Johnson</u>, United States

We did not include MODREX in our study. I am unable to
comment meaningfully on the difference between silo storage costs as
set forth in our study and costs associated with MODREX inasmuch as
the latter information has not previously been available to us.

<u>H. Konvicka</u>, Austria

Your slide on "technical evaluation matrix" showed a rather
low weighting for quality assurance. Furthermore, safety and
safeguards considerations were presented together with a rather high
weighting. Could you please elaborate a little bit on the reasons
for the different weightings and for tying safety and safeguards
together in one criterion ?

<u>E.R. Johnson</u>, United States

Quality assurance was subdivided into two components;
controllability of the process and ease of accomplishing Q.A., each
having 2 % of the weighting. We could really justify more than this
for the storage methods involved.

Safety and safeguards were subdivided into three components;
migration of radionuclides (8 %), control of contamination (8 %) and
ease of material control and accounting (4 %). All of these are
considered regulatory-related matters.

It should be pointed out that the weightings involved in the
evaluation are not absolute but rather have been established in such
a manner that differences in the methods can be clearly seen.

<u>H. Konvicka</u>, Austria

In the case of dry storage of FBR fuel, do you remove
residual sodium from the cladding surface, and if so, what is the
procedure ?

A. Sugier, France

Lorsque les assemblages sortis du réacteur ne retournent plus en sodium, ils sont alors lavés pour être stockés ou transportés.

Le procédé de lavage consiste à entraîner progressivement un brouillard d'eau dans un flux de gaz carbonique. Puis, en augmentant la quantité d'eau et en diminuant celle du gaz on lave progressivement et complètement en eau.

A. Uriarte, Spain

The SS basket you described for the Tn 1300 with 21 PWR and 1 kW/assembly. Is it a block or is it open ?

R. Christ, Federal Republic of Germany

It is not a block. It is an arrangement of 21 square channels.

B. Vriesema, The Netherlands

The neutron shield consists of tubes with fins in between. Can you comment on the fact that these fins form neutron leakages.

Second, could you comment on the possibility of actual leakage of tubes followed by loss of water.

P.G.K. Doroszlai, Switzerland

The tubes with the moderating water and boron acid neutron poison inside them have some effect as a collimator on the escaping high energy neutrons. So most of the neutrons are directed into the fins with the corresponding shielding effect. There is, however, a corresponding peak value of neutron radiation between the tubes which was considered in the shielding analysis.

The tubes have air-filled expansion hoses included. These hoses are provided with air pressure check valves. The water content may be controlled periodically by this means.

In case of accident (drop), partial loss of neutron shielding capability has been allowed for.

G. Pauluis, Belgium

La capacité de TOR ne suffisant pas au retraitement à la fois de Phénix et Superphénix, pourriez-vous nous dire quelles sont les options considérées pour le stockage à moyen terme des éléments Superphénix, sachant que dans le concept de stockage pour TOR, vous pouvez reprendre le combustible au bout de quelques mois, et que Monsieur Christ a mentionné une étude parallèle de stockage à sec en château ?

Quelles sont les options retenues par le CEA et un choix définitif a-t-il été arrêté ?

A. Sugier, France

L'option retenue consiste à laisser refroidir le combustible en stockage sur le site jusqu'à ce que la puissance soit compatible avec un transport à sec (type PWR). Pour le stockage intérimaire sur le site, deux voies sont étudiées en parallèle en vue d'un choix prochain :

- stockage d'un assemblage en étui individuel avec sodium et mise
en place des étuis dans un château de stockage (7 étuis/château);
- stockage direct en piscine après lavage.

Ensuite, après transport à sec, les combustibles seront stockés en
piscine en attente de leur retraitement.

J. Puga, Spain

In the 21 fuel assemblies design, you said that heat
convection is important. Which gas are you using to fill the cask ?

R. Christ, Federal Republic of Germany

It is not yet decided. But an inert gas.

J. Puga, Spain

But in any case not air ?

R. Christ, Federal Republic of Germany

We have not yet fixed the type of filler gas. But in any
case it will not be air, it will be an inert gas.

J. Geffroy, France

Les châteaux de stockage n'ont pas toujours d'amortisseurs
de chocs. Dans ce cas quel est le critère de décélération maximale
à ne pas dépasser pour une bonne tenue du combustible durant les
manutentions du château de stockage.

C.J. Ospina, Switzerland

The shock absorbers are required only during transportation
and handling of the cask. During storage, the cask itself is able,
from design and tests, to take care of the fuel under extreme external
forces e.g. earthquake, missile impact, etc. The design criteria are
such, that any component under any circumstance can withstand very
high deceleration limits e.g., up to 350 g.

H.D.K. Codée, The Netherlands

Did you perform a dose assessment for the four alternatives
you studied ?

E.R. Johnson, United States

No. We did, however, consider the prospective radiation
doses in assessing the licensability of the method.

In the case of the storage cask we assumed a radiation level
of 20 mr/hr at the surface, but we did not consider the effects of
skyshine, and that should be further analyzed.

P. Véron, Spain

What was the effect of the fire test on the damper ? Will the damper work properly in the case of a fall after a fire ?

R. Christ, Federal Republic of Germany

The shock absorber consists of wood completely enclosed in steel sheet. There was only minor charring of the wood. This shock absorber would remain functional when subjected to drop after a fire.

SESSION 3

Chairman - Président

E.O. MAXWELL

(United Kingdom)

SEANCE 3

DRY STORAGE INSTRUMENTED BUNDLE EXPERIMENTS

G.Kaspar, M.Peehs
Kraftwerk Union Aktiengesellschaft
P.O. Box 3220, D-8520 Erlangen/FRG

J.Fleisch
Deutsche Gesellschaft für Wiederaufarbeitung
von Kernbrennstoffen mbH
P.O.Box 1407, D-3000 Hannover/FRG

H.Unger
Kernkraftwerk Obrigheim GmbH
P.O.Box 1000, D-6951 Obrigheim

ABSTRACT

Analysis of the potential degradation mechanisms for spent LWR fuel under dry storage indicates that hoop strain would appear to limit the insertion temperature to 400 °C. To demonstrate the storage performance, two tests are conducted at ϑ_{max} = 300 °C (2 kW decay heat) and ϑ_{max} = 400 °C (3 kW decay heat). The fuel elements are composed of different design and burnup up to an average rod burnup of more than 40 GWd/tU accumulated in 4 reactor cycles. The experiments are instrumented with thermocouples and equipped with a gas monitoring line. Experiment No. 1 (ϑ_{max} = 300 °C) was terminated after 60 days. Experiment No. 2 (ϑ_{max} = 400 °C) is still in operation. There is no indication of any defect.

1. INTRODUCTION

The wet storage of spent LWR fuel, including the use of high density storage racks can be regarded as a proven technology. At present, increasing attention is being focussed on dry storage in the expectation that these facilities can be built and operated at lower costs than wet facilities, while providing the high degree of reliability required. Dry storage of spent fuel has advanced in the areas of both material characterization and operational experience /¯1 - 4_7 to the extent that it be considered a viable technology. Several countries are in advanced stages of pilot size demonstration of various dry storage techniques. A license application has also been submitted in the Federal Republic of Germany for storage in CASTOR casks. For the AFR-facility of 1500 t capacity at Gorleben the first and second construction license has been issued.

The only two important design criteria for a dry storage facility /¯1,2,3_7 are the choice of the storage medium and the fixation of the maximum cladding temperature. The storage medium and especially the presence of oxygen are of particular concern. Storage temperature is important because of the fact that cladding degradation mechanisms /¯2,4_7 will be initiated or acclerated by higher temperatures.

The following objectives for a dry storage experiment with instrumented spent fuel bundles may be derived on the basis of the considerations discussed:

- demonstration of the storage performance at the theoretically predicted upper temperature limit

- demonstration of the storage performance of higher burnup spent fuel bundles including fuel rods unloaded after 4 reactor cycles

- verification that no important degradation mechanisms have been overlooked.

2. ASSESSMENT OF THE POTENTIAL DEGRADATION MECHANISMS

Figure 1 features a compilation of all noteworthy degradation mechanisms. Oxidative corrosion caused by the impurities of technical grade inert gases used as storage medium may be neglected. The same is true for the hydrogen pick-up from the residual moisture. The maximum local hydrogen concentration due to thermodiffusion is much lower than the hydrogen concentration picked up during reactor operation. Fuel rod internal fission product corrosion may be excluded since no further fission products are released during extended storage /¯5_7. Crack propagation does not occur in the case of crack sizes smaller than 300 μm and temperatures less than 450 °C. Only the hoop strain limits the maximum insertion temperature to $\vartheta = 400$ °C, if the total strain is limited to 1 % uniform elongation. Both the morphology and the adherence of oxide layer and crud deposits may not be altered.

To achieve the abjectives stated above, the experiments are performed at $\vartheta_{max} = 300$ °C and $\vartheta_{max} = 400$ °C in N_2 from 2 bar at a water vapour pressure in equilibirum with the pool water of 40 °C. The average rod burnup within the test bundles varies between 22 GWd/tU and 43 GWd/tU. Each assembly contains in its central region fuel rods that have been in service for 4 reactor cycles.

3.   EXPERIMENTAL WORK

        The experimental work is being performed in the storage pool
of the Obrigheim reactor. Test fuel assemblies are used which have
a cross-like cut in their head to allow ease of fuel rod documentation and inspection (Fig. 2), 13 thermocouples are integrated in 3
rods inserted in the control rod guide tubes. The redesigned and specially equipped dry sipping box (Fig. 3) is used as the test bed. The
lid of the sipping can is penetrated by 3 tubes containing the connections for the thermocouples and a gas pipe for N_2 and gas sampling
during the test. Fig. 4 shows the test bed arrangement in the storage
pool.

        The characteristic features of both fuel bundles were analy-
zed by measuring selected fuel rods to determine the geometrical outer
profile, the eddy current characteristic and the oxide thickness
using KWU standard pool inspection equipment.

4.   EXPERIMENTAL RESULTS

4.1   THE 300 °C TEST

        Experiment No. 1 was performed with a fuel bundle with a
10 month decay time after reactor shutdown. At the start of the experiment the total decay heat of the assembly was 2 kW. As predicted,
the maximum bundle temperature reached nearly 300 °C. Figure 5 shows
the axial fuel rod temperature profile in terms of relative units.
Figure 6 shows the temperature/time pattern during the 60 days test
duration. During the test fuel bundle was flooded with water several
times and set dry without any adverse effect on integrity.
The test was terminated by rapid flooding of the storage canister.
The maximum bundle temperature dropped within 2 minutes from 210 °C
to 70 °C (Fig. 7). A careful post-test sipping test verified the
integrity of the fuel bundle.

4.2   THE 400 °C TEST

        Experiment No. 2 was performed with a test bundle with only
a 4.5 month decay period after reactor shutdown. The decay heat at
the beginning of the test was 3 kW. The maximum bundle temperature
was predicted to be 400 °C. The prediction was based on a computer
code calibrated with electrically heated bundles /⁻6_7 and by the
test number 1. The temperature measurement at the beginning of the
experiment verified satisfactorily the predicted temperatures as
shown in Fig. 6. The relative temperature distribution was in accordance with the experience from test No. 1.

        The temperature drop against time is plotted in Figure 6.
The measurement corresponds to the decrease of decay heat with time.
To check the integrity of the stored spent fuel assembly, gas samples
were taken several times. Up to now there is no indication of a defect in the fuel bundle. The test is still in progress (May 1982).

5.   DISCUSSION OF THE EXPERIMENTAL RESULTS

        Figure 8 summarizes the actual world-wide status of expe-
rience in experimental dry storage of spent LWR-fuel. Within the test
series the highest temperatures under dry storage conditions have
been reached with complete fuel bundles. The maximum temperatures are
only exceeded by single rod storage in the hot cells of Battelle
Columbus Laboratories. Besides the maximum bundle temperature, the
burnup of the stored spent fuel is also of great interest. Figure 9
shows the world-wide dry storage experience evaluated by burnup over

storage time. The viewgraph shows that an increasing burnup has no adverse effect on dry storage performance. Both fuel bundle had at the beginning of the tests a decay time of less than one year. These are the shortest decay periods under dry storage conditions ever demonstrated. This shows that the prestorage period is not a harmful parameter in terms of cladding degradation.

Assessment of the dry storage performance of spent LWR assemblies indicates that–under consideration of the temperature drop with time - the storage period is not the limiting parameter. On the contrary the maximum insertion temperature is the governing design criterion. Figure 10 demonstrates this fact. It shows the calculated uniform hopp strain for a representative temperature/time characteristic. Consequently the performed tests demonstrate the feasibility of dry storage under the most conservative circumstances. This is not only true of the temperature, it is also true of the oxidation potential of the storage atmosphere, since it contains a constant partial water pressure of about 50 mbar in the N_2.

After termination of tests No. 2, the precharacterized fuel rods will be recharacterized. These measurements will be performed to provide the data basis for the verification of the models used to predict the dry storage performance of spent LWR fuel.

6.   SUMMARY

To enhance the data bases for dry storage of spent LWR fuel and to verify the long-term behaviour in the temperature range up to ϑ = 400 °C, instrumented experiments with complete KWO fuel bundles are performed. The fuel bundles are rods of different design and burnup up to an average rod burnup of more than 40 GWd/tU accumulated in 4 reactor cycles. The experiments are performed in a sepcially designed dry storage box installed in the pool of the Obrigheim power reactor (KWO). To record continuously the cladding temperature, the test bundles are instrumented with 13 thermocouples each. At any time samples may be taken from the storage box to prove that the fuel bundle stays free of any defect.

Experiment No. 1 was performed with a fuel bundle with a 10 month decay time generating about 2 kW decay heat. The maximum bundle temperature reached 300 °C. Terminating this test run after 60 days, the maximum bundle temperature had decreased to 270 °C. The gas samples taken indicated that the fuel bundle was fully intact. Experiment No. 2 used a fuel bundle with only 4,5 months decay time after reactor shutdown. Upon starting the experiment, the heat generation of the test bundle was about 3 kW. The maximum bundle temperature reached 400 °C. Up to now there is no indication of any defect (until May 82).

Both fuel bundels were pre-characterized by measuring selected fuel rods (profilometry, eddy-current, oxide-thickness) prior to the experiments. Post test examination - especially of the test bundle from experiment No. 2 - will be performed to verify the predicted storage behaviour. These instrumented experiments together with post-test examination will be a useful contribution to the available data base for dry storage behaviour of spent LWR fuel.

7. REFERENCES

/⁻1⁻7   Proceedings of an IAEA advisory group/specialist meeting on
         "Spent Fuel Storage Alternatives", Las Vegas, USA, November
         1980, Summary and observations, p. 3 trough 19

/⁻2⁻7   M.Peehs, G.Kaspar, W.Jung and F.Schlemmer
         "Long Term Storage Bahaviour of Spent LWR Fuel" PATRAM 80,
         November 80, Berlin, FRG, Proceedings Volume II, p.393-947

/⁻3⁻7   K.Einfeld
         "Interim Dry Storage of Spent Fuel in Transport Casks",
         Proceedings of an IAEA advisory group/meeting on "Spent
         Fuel Storage Alternatives", Las Vegas, USA, November 1980,
         Summary and observation, p.443 through 444

/⁻4⁻7   C.R.Blomgren,
         "Summary of Spent Fuel Dry Storage Testing at the E-MAD
         Facility", Proceedings of an IAEA advisory group/specialist
         meeting on "Spent Fuel Storage Alternatives", Las Vegas,USA,
         November 1980, Summary and observation, p. 257 through 325

/⁻5⁻7   G.Kaspar, W.Haas, M.Peehs, E.Haas, M.Beuerle
         Jod-Exhalation aus UO_2 unter stationären und transienten
         Bedingungen"
         Abschlußbericht Förderungsvorhaben BMFT RS 285, 1981.

Figure 1   **Kraftwerk Union**

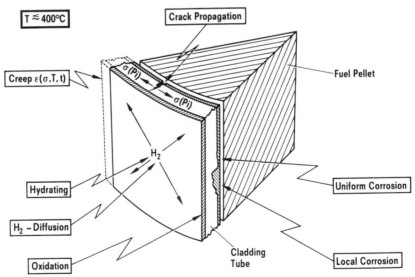

Mechanisms Affecting Spent Fuel Cladding Performance during Dry Storage

E 82 420 e

- 118 -

Kraftwerk Union

Figure 2

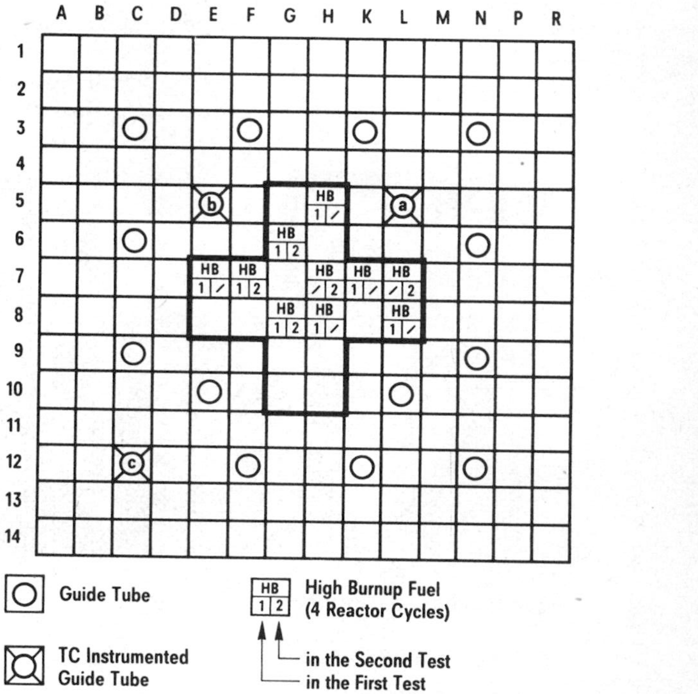

Location of TC Instrumented Guide Tubes and High Burnup Rods in the KWO Dry Storage Test Assemblies

Kraftwerk Union

Figure 3

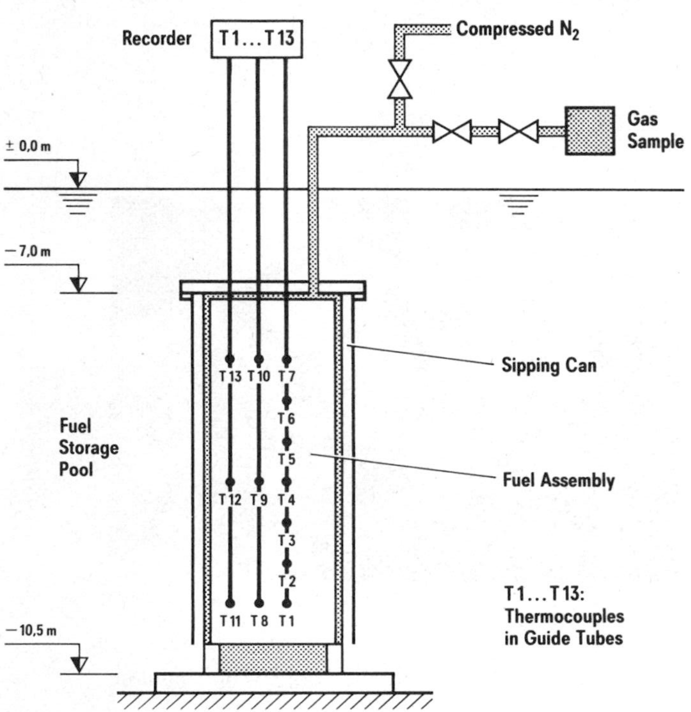

Scheme of KWU Dry Storage Test Device at KWO

Figure 4

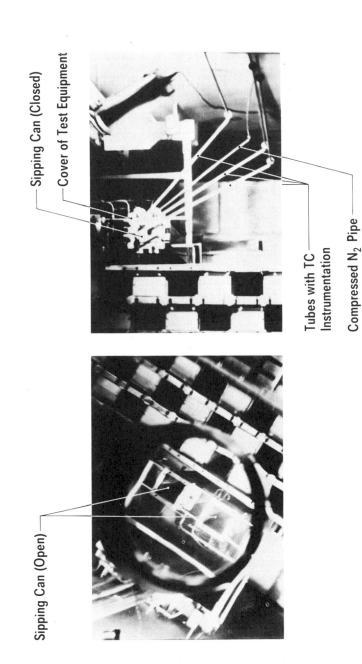

Sipping Can (Open)

Sipping Can (Closed)

Cover of Test Equipment

Tubes with TC Instrumentation

Compressed N_2 Pipe

Dry Storage Test Equipment in the KWO Spent Fuel Pool

Figure 5   *Kraftwerk Union*

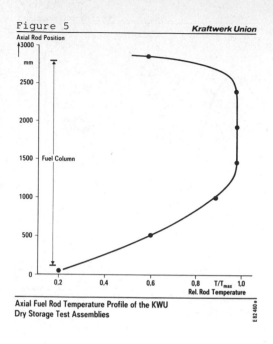

Axial Fuel Rod Temperature Profile of the KWU Dry Storage Test Assemblies

E 82 460 e

Figure 6   *Kraftwerk Union*

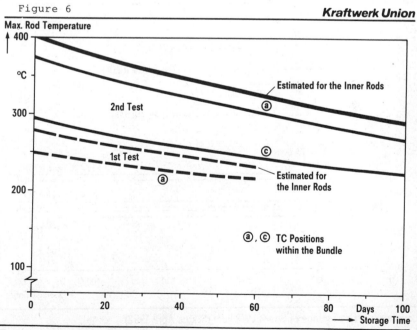

Temperature History of KWU Dry Storage Tests during the First 100 Days of Storage

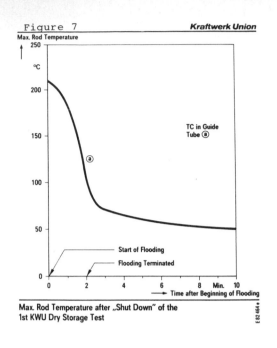

Figure 7   **Kraftwerk Union**

Max. Rod Temperature after „Shut Down" of the
1st KWU Dry Storage Test

Figure 8   **Kraftwerk Union**

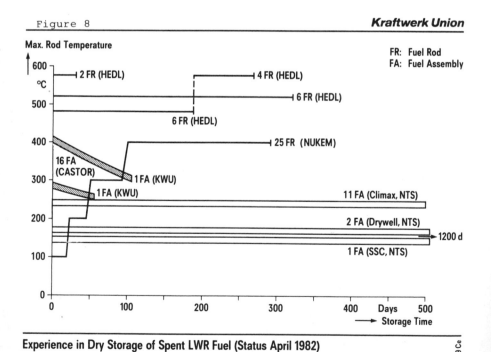

Experience in Dry Storage of Spent LWR Fuel (Status April 1982)

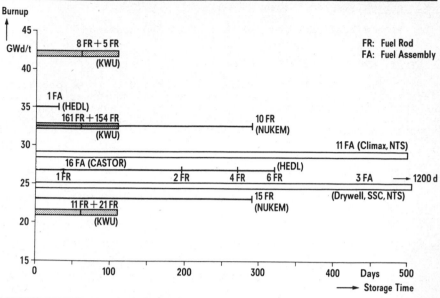

Experience in Dry Storage of Spent LWR Fuel (Status April 1982)

E 82 428 Ce

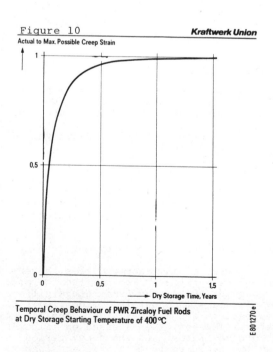

Temporal Creep Behaviour of PWR Zircaloy Fuel Rods
at Dry Storage Starting Temperature of 400 °C

E 801270 e

DISCUSSION

A.B. Johnson, United States

      How long will your 400°C test at KWO continue ?

M. Peehs, Federal Republic of Germany

      The test will run through the year 1982. if we decide to
start a third experiment - and there are plans now under discussion -
the present experiment will be terminated on the availability of the
next test element. Since a third experiment should cover a higher
temperature regime a possible experiment 3 should start in the fall,
1982.

S.J. Naqvi, Canada

      Are your planning any destructive tests in your experiment 2.
If so, what are they ?

M. Peehs, Federal Republic of Germany

      After termination of the test we will perform a careful pool
inspection. We will measure the geometrical profile, oxide thickness
and eddy current characteristics. We do not expect to detect any
adverse effects. However, if that is not the case we will transfer
the fuel rod in question to the hot cell. If we do not find such
effects, we will select one rod being typical for hot cell inspection.

R.J. Pearce, United Kingdom

      Did you say you had used different clad materials in these
tests and, if so, what were they ?

M. Peehs, Federal Republic of Germany

      Since we have reactor fuel rods from irradiation testing in
the KWO available with different cladding, we also inserted in the
test bundle zircalloy clad with different heat treatment in the
central region as well as four cycle fuel rods. This way we accumulate
not only experience on fuel of standard design. We also get
information on advanced fuel.

J. Geffroy, France

      Dans vos expériences sur assemblages irradiés, que pensez-vous
des incertitudes sur les mesures de température et de puissance ?

M. Peehs, Federal Republic of Germany

      The accuracy of the thermocouples is very good since we did
a lot of work on calibration. Because the thermocouples are inserted
in control rod guide tubes, the temperature difference to the
neighbouring claddings is about 10°C (too low). So in any case the
measured temperatures are conservative. The uncertainty of the decay
heat is greater. The estimate is based on calculation taking into
account the burn up history and the decay period. Since the
predicted and the measured temperatures are in a good accordance, the
input data on decay heat seems to be fairly good.

SPENT FUEL BEHAVIOR IN DRY STORAGE

A. B. Johnson, Jr.
P.A. Pankaskie
E.R. Gilbert

ABSTRACT

Dry storage is emerging as an attractive and timely alternative to complement wet storage and assist utilities to meet interim storage needs. Spent
fuel has been handled and stored under dry conditions, at least for brief
periods, almost from the beginning of nuclear reactor operation. Dry well and
vault facilities have been used to store certain types of fuel since 1964; and
more recently, programs have developed to demonstrate dry storage of irradiated
Zircaloy-clad fuel in metal casks, dry wells, silos, and vaults. Hot cell and
laboratory studies are also under way to investigate specific phenomena related
to cladding behavior in dry storage.

A substantial fraction of the U.S. light-water reactor spent fuel inventory has aged for relatively long times and has relatively low decay heats,
which suggests that much of the U.S. fuel inventory can be stored at relatively
low temperatures. Alternatively, rod consolidation of older fuel can be considered without exceeding maximum cladding temperatures.

As interim dry storage concepts develop, consideration should be given,
where practical, to measures that will simplify future fuel management operations, i.e., reprocessing and/or geologic disposal.

Work Supported by the
U.S. Department of Energy
Under Contract DE-AC06-76RLO 1830

Pacific Northwest Laboratory
Richland, Washington 99352

INTRODUCTION

At the end of 1980 there were approximately 28,000 light-water reactor (LWR) fuel assemblies in interim storage in U.S. spent fuel pools:[1] 18,000 boiling water reactor (BWR) and 10,000 pressurized water reactor (PWR) assemblies. By 1990 the total number of U.S. LWR assemblies is estimated to rise to approximately 150,000.[a] It is currently unlikely that the number of stored U.S. assemblies will decrease significantly over the next decade due to reprocessing and/or permanent storage in geologic repositories.

By 1986, seven U.S. reactors are expected to fill current wet storage capacities and will need additional storage space or face the prospect of ceasing operation. By 1990, a total of 29 U.S. reactors are expected to expend current storage space. Several options are being evaluated to meet future interim storage needs:

- additional wet storage facilities
- rod consolidation
- double tiering
- dry storage.

Additional wet storage facilities may require ∿6 years to license and construct.[2] Rod consolidation has not yet been licensed or demonstrated with irradiated fuel in the United States, although there is strong interest in the concept; and not all U.S. pools could implement rod consolidation due to floor loading considerations without special provisions. Double tiering is being implemented at two U.S. pools, but it is not being considered for large-scale development. Dry storage, however, is emerging as a technically viable, cost-effective option for at-reactor (AR) interim storage,[2] and technical and economic comparisons are available for dry and wet away-from-reactor (AFR) storage.[3] Furthermore, development of two dry storage AFR sites is under way in the Federal Republic of Germany (FRG). The period from licensing to availability for dry storage appears to be compatible with 1986 storage needs, provided that licensing developments are timely.

This paper summarizes the experience base for dry storage of spent nuclear fuel and the status of technical issues relating to fuel behavior in dry storage.

(a) This number is subject to revisions due to reactor cancellations and deferrals.

HISTORICAL ASPECTS OF IRRADIATED FUEL HANDLING UNDER DRY CONDITIONS

Irradiated nuclear fuel has been handled under dry conditions since the beginning of irradiated fuel examinations in hot cells in the mid-1940s, and irradiated water reactor fuel has been examined dry in hot cells since ∿1960. More recently, hundreds of water reactor fuel assemblies have been shipped under dry conditions. Zircaloy-clad fuel has resided in dry casks during sea transport for 2 to 3 months at estimated temperatures from 150 to 300°C[a]-- the range where much dry interim storage of irradiated fuel is expected to occur. Some irradiated water reactor fuel rods have been stored continuously at hot cells under dry conditions for approximately a decade, although at relatively low temperatures (<100°C). While there have been no cases where Zircaloy-clad fuel has been observed to degrade under dry conditions, bases more relevant to dry interim storage are being developed.

BASES FOR DRY INTERIM STORAGE OF IRRADIATED FUEL

Four dry storage concepts are being demonstrated with irradiated fuel: metal cask, dry well, silo, and vault. The following major elements contribute to the developing dry storage data base:

- Gas-cooled reactor (GCR) and liquid metal fast breeder (LMFBR) fuel has been stored dry in vaults and dry wells.

- Dry storage demonstrations of Zircaloy-clad fuel are being conducted in cask, dry well, silo, and vault facilities.

- Hot cell and laboratory investigations of materials used in water reactor fuel assemblies and facilities are being conducted.

- Published literature exists that is relevant to water reactor fuel behavior under dry conditions.

SIGNIFICANCE OF GCR AND LMFBR DRY STORAGE EXPERIENCE

Table 1 summarizes dry storage experience for GCR and LMFBR fuel[4] that is relevant to interim dry storage of other fuel types in the following areas:

- development of safety analyses for dry storage facilities

- demonstration of successful emplacement and retrieval of irra- diated nuclear fuel in dry well and vault facilities

---

(a) Fuel cladding temperatures up to 450°C have been estimated for some LWR fuel shipped dry to La Hague, with no evidence that either shipment condi- tions or fuel handling caused cladding failure.

- demonstration of safe operation of dry storage facilities over extended periods (dry well and vault since 1964)

- demonstration of low radiation levels and occupational exposures for dry well and vault facilities.

TABLE 1. Dry Storage of GCR and LMFBR Fuel[a]

Site	Storage Facility	Facility Type	Fuel Type/Cladding	Earliest Fuel Storage	Cover Gas
INEL	IFSF	Vault	HTGR/Graphite	1975	Air
INEL	HFEF/S	Vault	LMFBR/SS	1964	Argon/Air
INEL	HFEF/N	Vault	LMFBR/SS	1975	Argon
Wylfa	Wylfa	Vault	Magnox/Mg Alloy	1971	Air/CO_2
Wylfa	Wylfa	Vault	Magnox/Mg Alloy	1979	Air/CO_2
INEL	ICPP	Dry Well	HTGR/Graphite	1971	Helium
INEL	ICPP	Dry Well	LMFBR/SS	1974	Helium
INEL	RSWF	Dry Well	LMFBR/SS	1964	Air/Argon

(a) See glossary for definition of terms.

DRY STORAGE DEMONSTRATIONS: ZIRCALOY-CLAD FUEL

Table 2 summarizes dry storage demonstrations of irradiated Zircaloy-clad fuel assemblies in cask, dry well, silo, and vault facilities.[4-7] A schematic view of a metal storage cask is shown in Figure 1; Figures 2 through 6 are photographs of actual dry storage facilities and operations involving Zircaloy-clad fuel. The ongoing demonstrations provide the following insights to interim dry storage of Zircaloy-clad fuel:

- Fuel has been emplaced in cask, dry well, silo, and vault facilities and retrieved from dry well and vault facilities. Fuel retrieval from cask and silo demonstrations is planned.

- Safe operation and storage has been demonstrated for all four concepts to date.

- Canistered fuel has been stored in dry well, silo, and vault facilities; retrieval from canisters has also been demonstrated.

- Low radiation levels in personnel access areas and low occupational doses during fuel handling and storage have been maintained.

TABLE 2. Dry Storage Demonstrations of Zircaloy-Clad Fuel[a]

Location	Storage Facility	Facility Type	Fuel Type	Earliest Fuel Storage	Number of Assemblies
Nevada/USA	E-MAD	Surface Dry Well	PWR	1979	4
Nevada/USA	Climax	Deep Dry Well	PWR	1980	11
Nevada/USA	E-MAD	Silo	PWR	1978	1
Nevada/USA	E-MAD	Vault	PWR	1978	1[b]
Manitoba/Canada	WNRE	Silo	WR-1	1975	138[c]
Manitoba/Canada	WNRE	Silo	CANDU	1976	360[c]
FRG	Wuergassen	Metal Cask	BWR	1982	16

(a) See glossary for definition of terms.
(b) Used for temporary storage of 17 PWR assemblies.
(c) Assemblies ∿0.5 m in length, compared to 4 to 4.5 m for LWR assemblies.

Several types of water reactor fuel are included in the demonstrations: BWR fuel is being stored in FRG cask tests; pressurized heavy-water reactor (PHWR) fuel, in Canadian silo tests; and PWR fuel, in U.S. dry well, silo, and vault tests.

To date there has been no evidence that fuel degradation has occurred during dry shipment, handling, or storage. At the Nevada Test Site (NTS) E-MAD[a] facility, canister cover gases were sampled after periods of dry storage for evidence (principally ^{85}Kr) that the Zircaloy cladding had failed.[8] Seven canisters were sampled, and none showed detectable ^{85}Kr. Four of the canisters had been in vault storage for an average of 5 months at temperatures of ∿230°C. One canister was exposed in the fuel temperature test to a range of temperatures, with a brief maximum of 300°C. Two canisters were exposed for ∿18 months in dry wells from ∿180 down to 150°C. Detailed examinations of Zircaloy-clad fuel after extended dry storage remain to be conducted.

Metal casks are regarded as the leading candidate for interim dry storage. Dry storage cask demonstrations are being planned in the United States, and major AR and AFR dry cask storage facilities are being developed in the Federal Republic of Germany. Cask storage is also being implemented for dry interim storage of test reactor fuel in Switzerland.[9]

Time-temperature relationships for various dry storage demonstrations involving Zircaloy-clad fuel are compared in Figure 7. Hypothetical temperature decay curves for PWR assemblies residing in a cask at maximum initial temperatures of 250 and 400°C are also indicated. The decay curves are based on calculated thermal characteristics for a dry multielement cask and must be regarded as preliminary until actual cask temperature monitoring data are available.

(a) E-MAD--Engine-Maintenance Assembly and Disassembly.

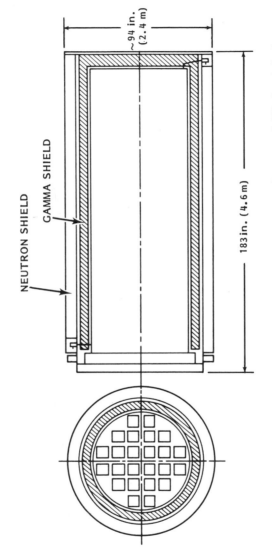

FIGURE 1. Schematic of Metal Storage Cask for Dry Storage of PWR Spent Fuel

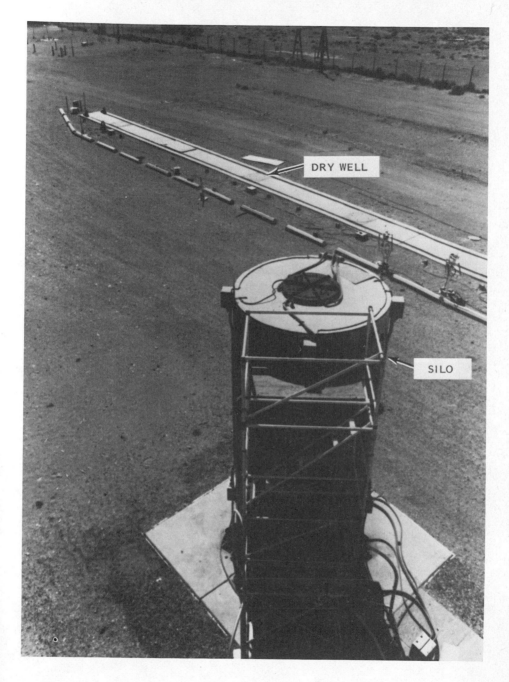

FIGURE 2. Four Surface Dry Wells, Each Storing One Irradiated PWR Assembly--E-MAD, Nevada Test Site, USA (fueled concrete silo shown in foreground)

SPENT FUEL TEST - CLIMAX GRANITE

425 m

FIGURE 3. Schematic of Underground Dry Wells in Climax Granite--Nevada Test Site, USA

FIGURE 4. Underground Dry Wells in Granite Chamber. Eleven irradiated PWR assemblies are stored in the dry wells.

FIGURE 5. Concrete Silo Storing One Irradiated PWR Assembly--E-MAD, Nevada Test Site, USA. Silo dimensions: height = 6.4 m; diameter = 2.6 m.

FIGURE 6. Main Hot Cell for Remote Receipt and Handling of Irradiated Fuel Assemblies--E-MAD, Nevada Test Site, USA. Irradiated assembly is inside stainless steel canister being transferred to transfer cask; storage vault is out of view to the left.

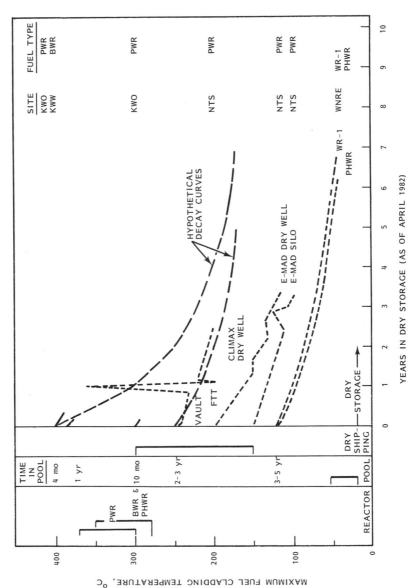

FIGURE 7. Fuel Assembly Temperature Histories: Dry Storage Demonstrations

The following observations can be drawn from Figure 7:

- No single demonstration covers the full range of expected temperature decay histories for assemblies that begin dry storage at 250 and 400°C although the demonstrations collectively cover most of both temperature regimes. FRG tests cover selected temperatures in the upper range (300 and 400°C); in fact, in the Wuergassen cask test, only a few central sections of rods were at 400°C. NTS tests cover the intermediate range (100 to 250°C with brief increases to 300 and 360°C on two assemblies); and Whiteshell Nuclear Research Establishment (WNRE) tests cover the lower temperature range (40 to 120°C).

- Several fuel types are represented: BWR, PHWR, PWR, and WR-1 (organic-cooled Whiteshell reactor); and fuel with Zircaloy-2 and Zircaloy-4 cladding is included.

- Comparisons to reactor cladding temperature ranges indicate that KWO (Obrigheim, PWR) and KWW (Wuergassen, BWR) tests are conducted above the reactor operating temperature range for fuel cladding for the fraction of rods that resided at the maximum temperature. NTS and WNRE tests are conducted below the reactor operating range except for brief temperature tests on one NTS assembly up to ∿360°C.

- The assemblies had cooling times varying from 1 year for KWW assemblies, to 2 to 3 years for NTS assemblies, to 3 to 5 years for WNRE assemblies.

- Fuel handling histories vary for the various demonstrations. The KWW and KWO demonstrations were conducted at the reactors where the fuel was discharged; the NTS assemblies were discharged at the Turkey Point reactor (PWR). Some Turkey Point assemblies were shipped dry to the Battelle Columbus Laboratories (BCL) for characterization; some were shipped dry directly to NTS from the reactor. The WR-1 fuel was discharged at the dry storage site (WNRE); the PHWR fuel was discharged from the Douglas Point reactor and shipped to the storage site. Temperatures during dry shipments are not known precisely but probably fall within the general range of 150 to 300°C depending on the assembly heat rating. Temperatures for assemblies stored dry in a hot cell probably are less than 100°C based on BCL observations.

- Zircaloy-clad fuel has been stored dry for nearly 7 years but at relatively low temperatures.

At 400°C, cladding radiation damage tends to anneal in extended exposures; therefore, cladding mechanical properties will change relative to as-irradiated properties. Ductilities will increase and strengths will decrease. On the other hand, cladding in demonstrations at 250°C and below will tend to maintain properties close to irradiated properties.

HOT CELL AND LABORATORY STUDIES

The preceding section summarized demonstrations that are under way on full Zircaloy-clad assemblies. Other studies (Table 3) are planned or under way using irradiated Zircaloy-clad rods or rod sections and unirradiated materials

TABLE 3. Dry Storage Studies--Hot Cell and Laboratory[a]

Country	Organization	Fuel Type
Canada	WNRE	PHWR
	Ontario Hydro	PHWR
FRG	DWK, KWU	BWR, PWR
	German Ministry of Res. & Tech. DWK, JRC, Ispra, KWO, KWU, PE, and TN	BWR, PWR
USA	DOE	BWR, PWR
	NRC	BWR, PWR
	TVA/EPRI	PWR and Unirradiated

(a) Scopes of most of the studies are still being defined. See glossary for definition of terms.

to define specific phenomena related to cladding and fuel pellet behavior in dry storage. Phenomena to be investigated are:

- prospects for stress rupture to occur during dry storage

- behavior of through-wall and incipient cladding defects under dry storage conditions

- prospects for stress corrosion cracking (SCC) to occur in dry storage regimes

- oxidation rates of fuel pellets exposed to an oxidizing cover gas at cladding defects

- cladding reactions with cover gases

- behavior of crud in various dry storage environments.

The hot cell and laboratory studies are directed to definition of specific phenomena. The demonstrations involve larger numbers of fuel rods and provide better statistics. Together, the two approaches have complementary value to development of a dry storage data base.

BASES FOR ASSESSING DRY STORAGE BEHAVIOR OF ZIRCALOY-CLAD FUEL

The dry storage demonstrations and hot cell and laboratory experiments that are planned or under way appear to address the range of data needs and cladding performance definition over the expected range of dry storage regimes. An earlier assessment identified the expected range of potential cladding degradation mechanisms:[10]

- stress rupture
- mechanical overload
- SCC
- fracture of flawed cladding
- delayed hydrogen cracking and fatigue
- internal hydriding
- oxidation (internal and external).

It was concluded that stress rupture was the most probable controlling mechanism for temperatures near 400°C; however, uncertainty was also expressed regarding the possible role of SCC. Subsequent testing of irradiated rods at 480 to 570°C resulted in annealing of radiation damage,[11] and no stress rupture occurred. However, there are reservations regarding extrapolation of the results to lower temperatures where radiation damage remains in the cladding.

Fuel and cladding oxidation is precluded in inert atmospheres, but there are concerns that substantial monitoring may be required although it may be necessary only until fuel temperatures drop to levels where oxidation is no longer a threat to cladding and fuel. In oxidizing atmospheres, Canadian data[12] suggest that fuel oxidation may contribute to cladding degradation if temperatures are sufficiently high to promote substantial conversion of UO_2 to U_3O_8, which results in a volume increase.

The fuel demonstrations and hot cell tests that are under way and being planned will expose a variety of fuels to a range of temperature regimes. In some cases the cover gases will be oxidizing; in some cases, inert. These demonstrations and tests will provide opportunities to observe those potential cladding failure mechanisms that are recognized and those that may not be well defined. Several studies are under way that will address expected degradation mechanisms. Fuel assemblies will be examined in Canadian, FRG, and U.S. demonstrations.

The possible effects of reactor crud should be considered during fuel retrieval and in decommissioning the dry storage facility. Crud behavior is being addressed in some of the dry storage studies, and opportunities exist to make observations on crud behavior in the demonstrations.

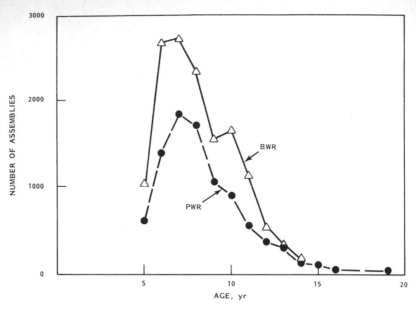

FIGURE 8. Distribution of Wet Storage Times for U.S. LWR Spent Fuel Inventories Discharged as of December 1980 and Projected Forward to 1986

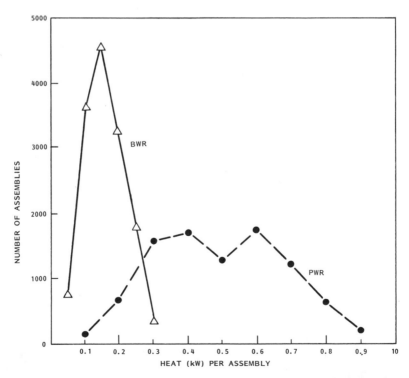

FIGURE 9. Distribution of Decay Heats for U.S. LWR Spent Fuel Inventories Discharged as of December 1980 and Projected Forward to 1986

Although dry storage facilities (casks, vaults, dry wells, silos, canisters) are not expected to degrade significantly during interim dry storage, systematic observations and reporting during the demonstrations will provide bases to: 1) develop assurances of satisfactory storage system behavior and 2) respond to questions from licensing boards.

WATER REACTOR FUEL ASSEMBLY CHARACTERISTICS

CLADDING TYPES

Approximately 95% of the current U.S. LWR inventory involves Zircaloy-clad fuel,[13] and the remaining U.S. LWR fuel is clad with stainless steel. There are scenarios that suggest that dry storage of stainless steel-clad fuel could occur, but the prospects currently appear to be low.

FUEL AGES AND DECAY HEATS

In several countries older LWR fuel inventories have been shipped to reprocessing plants; therefore, the remaining spent fuel tends to have relatively short cooling times and high decay heats. In contrast, since no U.S. LWR fuel has been reprocessed since 1971, a large fraction of the U.S. inventory has relatively long cooling times (see Figure 8) and low decay heats (see Figure 9).

The data plotted in Figures 8 and 9 were generated based on the U.S. spent fuel inventory in storage at the end of 1980. The inventory characteristics were projected forward to 1986 to provide a basis to assess the inventory at the time that some U.S. reactors begin to deplete current storage capacities. No fuel discharged after 1980 is included in the assessment, and the estimates of kilowatts per assembly are based on the ORIGEN code. Lower average decay heats per assembly for BWR fuel arise from lower uranium inventories per assembly: ∿190 kg for BWR fuel and ∿460 kg for PWR fuel. Burnups also tend to be lower for BWR fuel.

The U.S. fuel inventory characteristics provide a certain flexibility for implementing dry storage:

- PWR fuel with decay heats less than 0.65 kW may be storable in casks at temperatures below 200°C although this is a highly preliminary estimate that should be confirmed by instrumented multiassembly cask tests; corresponding estimates have not yet been made for BWR fuel.

- At relatively low temperatures, successful storage of the fuel in air appears to be a reasonable expectation. Fuel cladding defects are less likely to present storage problems in this temperature regime.

- The inventory with low heat ratings offers good prospects for rod consolidation for dry storage while still maintaining satisfactory storage temperatures.

Inspection of projected storage inventory characteristics for U.S. reactors indicates that all the reactors that will begin to deplete storage capacities in 1986 have substantial fuel inventories with decay heats of 0.5 kW or below. Dry storage at moderate temperatures is a good prospect for those reactors. All but one of the reactors have crane capacities of 100 tons or above to handle multiassembly dry storage casks.

CLADDING DEFECT CHARACTERISTICS AND SIGNIFICANCE

It is broadly recognized that a small fraction of the irradiated fuel rods develop through-wall cladding defects during reactor operation. The defects have minimal impacts on wet storage.[14] These defects vary from pinholes to narrow cracks, to larger cracks or openings, and to a few cases where rod sections are missing. Many assemblies with defective rods can be identified by visual inspection and by fuel sipping (isolation of an assembly in a chamber to determine if fission products are evolving). However, the sensitivity of current defect identification does not assure that all defective rods will be identified, particularly forassemblies stored for extended times in water. Defective rods no longer have internal gas pressures, thus hoop stresses are nil. In inert atmospheres where fuel oxidation is precluded, cladding defects are not likely to cause storage problems; however, in oxidizing atmospheres, temperature regimes need to be identified where fuel oxidation rates are low enough to preclude significant fuel oxidation at cladding defects.[12]

Some early generations of LWR fuel had higher fuel failure rates than more recent fuel generations, but early fuel generations had relatively low burnups.[13] The older fuel currently has relatively long pool residence and low heat generation rates. Therefore, cladding defects in older fuel may not significantly impact dry storage, even in oxidizing atmospheres. In fact, U.S. studies are planned to investigate fuel oxidation at cladding defects. Fuel defect studies are under way in Canada at 150°C in moist and dry air.[7]

INCIPIENT CLADDING DEFECTS[a]

Incipient defects are not easily detected; however, numerous metallurgical examinations of irradiated Zircaloy cladding suggest that incipient defects are not prevalent even though they have been observed on occasion. Studies are planned in the United States to investigate the behavior of actual and/or simulated incipient defects under dry storage conditions.

ROD INTERNAL PRESSURES

Although early LWR fuel was not pressurized, helium fill gas pressures for PWR rods were increased from 1 atm to current levels (20 to 30 atm at room temperature) in the early 1970s. Additional pressures from fission gases released from the UO_2 pellets may increase PWR internal rod pressures to maximums near 65 atm; values of ∿40 atm at assembly average burnups of 35,000 MWd/MTU are typical.[15]

BWR rod pressures were increased in the late 1970s from 1 to 3 atm at room temperature. Designs have been proposed for rod pressures of 6 atm at room temperature. For high burnup approaching 40,000 MWd/MTM, fission gases can increase total BWR rod pressures to maximums near 35 atm at room temperature; values of ∿14 atm at burnups of 28,000 to 30,000 MWd/MTU are typical.[15] Operational internal rod pressures are ∿2 times room temperature values for PWRs and ∿1.8 times higher for BWRs.

During operation, reactor operating pressures (∿150 atm for PWRs and ∿70 atm for BWRs) oppose the fuel rod internal pressures, negating much of the operational cladding hoop stresses. Reactor back pressures are not present in dry storage, thus cladding hoop stresses will tend to be higher than under most reactor operating conditions.

THE ROLE OF FUEL CANISTERS

Three of the four dry storage concepts involve fuel emplacement in the storage facility under dry conditions. To avoid possible contamination dispersal from crud or in case a fuel assembly is dropped during transport or loading, fuel is placed in canisters for dry loading.[2] In contrast, irradiated fuel is loaded into storage casks in spent fuel pools. The cask is then drained, dried, and sealed. Movement and siting of the fuel inside the cask precludes a clear requirement to can the fuel, unless rod consolidation is conducted. Any recommendations to can the fuel for cask storage must be weighed against the drawbacks, which include cost, possible cask space impacts, increased cladding temperature and more difficult temperature modeling, and possible decanning and canister disposal.

(a) Defects that extend part way through the wall of the cladding.

Canning for any dry storage concept should involve consideration of canister designs that will be, if possible, compatible with subsequent operations (reprocessing and/or geologic disposal).

INTEGRITY OF STORAGE FACILITIES

Tables 1 and 2 indicated the location and variety of facilities for dry storage of irradiated fuel, including metal casks, dry wells, silos, vaults, fuel canisters, fuel transfer equipment, and monitoring devices. The first demonstration of irradiated fuel storage in a metal cask has been under way only since March 1982. However, metal shipping casks have been subjected to irradiated fuel, alternate wetting and drying, and a large variety of atmospheric factors for up to three decades. Vault and dry well facilities have operated at the Idaho National Engineering Laboratory (INEL) since 1964. Concrete silos have existed at WNRE since 1975.

While no comprehensive assessment of facility condition has been performed, operators of three facilities at INEL and facility operators at E-MAD and WNRE were polled regarding the integrity of facilities and equipment. In all cases operations have proceeded smoothly. There is currently little evidence that facilities and equipment have degraded substantially. Steel dry well liners in contact with soil have expected lifetimes of ~25 years without cathodic protection, and dry wells protected from the soil by concrete grout have expected lifetimes of 100 years.[4]

CONCLUSIONS

1. Experience with dry storage, handling, and shipping of spent fuel provides valuable support for safe interim dry storage of LWR spent fuel until reprocessing or terminal storage is provided.

2. Dry storage demonstrations and hot cell/laboratory experiments are planned or under way that address the range of data needs and cladding performance definition over the expected range of dry storage regimes.

3. Much of the U.S. LWR spent fuel inventory has been in pool storage long enough that the decay heat will generate relatively low peak temperatures during dry storage. These low temperatures will likely enable fuel rods to be stored in air environments with acceptably low levels of degradation; alternatively, rod consolidation is a good prospect for fuel with low decay heats.

4. Studies that are planned or under way to characterize potential fuel cladding degradation mechanisms such as fuel oxidation, SCC, stress rupture, and fuel cladding defect behavior will enable dry storage temperature limits to be established.

5. Although crud behavior is not expected to have a major impact on dry storage operations, it is being addressed in technology studies to determine possible effects on fuel retrieval and storage system decommissioning.

6. Storage system integrity is not expected to present major problems, but it needs to be addressed in technology studies.

7. Interim storage developments should consider, where practical, measures that will simplify future fuel management operations, (reprocessing and/or geologic disposal).

ACKNOWLEDGMENTS

The authors are grateful to A. R. Hakl of Westinghouse-Nevada for supplying photographs of NTS dry storage facilities; to R. J. Bahorich of Westinghouse AESD and L. B. Ballou of Lawrence Livermore National Laboratory for discussions regarding NTS fuel temperature histories; to M. Peehs of Kraftwerk Union, B. R. Teer of Transnuclear, and J. D. Rollins of GNS for pertinent discussions; to B. M. Cole of PNL for providing fuel inventory assessments; and to R. A. McCann of PNL for assessing fuel temperature behavior in dry storage. This work was sponsored by the U.S. Department of Energy (DOE), Commercial Spent Fuel Management Program, under Contract DE-AC06-76RLO 1830.

REFERENCES

1. Oak Ridge National Laboratory. August 1980. Spent Fuel and Waste Inventories and Projections. ORO-778, Oak Ridge, Tennessee.

2. Johnson, E. R., Associates. November 1981. A Preliminary Assessment of Alternative Dry Storage Methods for the Storage of Commercial Spent Nuclear Fuel. DOE/ET/47929-1.

3. Lawrence, M. J., G. R. Moore, and R. C. Winders. 1982. "Cost for Spent Fuel Management--An IAEA Study." Presented at International ENS/ANS Conference--New Directions in Nuclear Energy with Emphasis on Fuel Cycles--Brussels, Belgium, April 26-30, 1982. Published in ANS Trans. 40:140-142.

4. Anderson, P. A., and H. S. Meyer. April 1980. Dry Storage of Spent Nuclear Fuel. NUREG/CR-1223, Exxon Nuclear Co., Inc., Idaho Falls, Idaho.

5. Hakl, A. R. 1980. "Remote Handling Capabilities of E-MAD for Dry, Retrievable, Interim Storage of Spent Nuclear Fuel." In Spent Fuel Storage Alternatives, DOE-SR-0009, pp. 225-255.

6. Wright, J. B. 1981. "Spent Fuel Dry Storage--A Look at the Past, Present and Future." Presented at Fuel Cycle Conference '81, March 15-18, 1981, Los Angeles, California.

7. Ohta, M. M. 1980. "Status of Dry Storage of Irradiated Fuel in Canada." In Spent Fuel Storage Alternatives, DOE-SR-0009, pp. 383-408.

8. Bolmgren, C. R. 1980. "Summary of Spent Fuel Dry Storage Testing at the E-MAD Facility." In Spent Fuel Storage Alternatives, DOE-SR-0009, pp. 257-236.

9. Ospina, C. J. 1980. "Technical Overview of Intermediate Dry Storage of Spent Fuel." In Spent Fuel Storage Alternatives, DOE-SR-0009, pp. 457-512.

10. Blackburn, L. D., et al. 1978. Maximum Allowable Temperature for Storage of Spent Nuclear Reactor Fuel. HEDL-TME 78-37, Hanford Engineering Development Laboratory, Richland, Washington.

11. Einziger, R. E., S. D. Atkin, D. E. Stellrecht, and V. Pasupathi. April 1982. "High-Temperature Postirradiation Materials Performance of Spent Pressurized Water Reactor Fuel Rods Under Dry Storage Conditions." Nucl. Tech. 57:65-80.

12. Boase, D. G., and T. T. Vandergraaf. 1977. "The Canadian Spent Fuel Storage Canister." Nucl. Tech. 32:60-71.

13. Johnson, A. B., Jr., et al. May 1980. Annual Report FY 1979, Spent Fuel and Fuel Pool Component Integrity. PNL-3171, Pacific Northwest Laboratory, Richland, Washington.

14. Johnson, A. B., Jr. 1978. "Impacts of Reactor-Induced Defects on Spent Fuel Storage." In Proceedings of Spent Fuel Elements, Madrid, Spain, June 20-23, 1978.

15. Beyer, C. E. 1982. "An Evaluation of Published High-Burnup Fission Gas Release Data." Paper presented at the American Nuclear Society Topical Meeting on Fuel Performance and Utilization, April 4-8, 1982, Williamsburg, Virginia.

GLOSSARY AND ABBREVIATIONS

AFR--away from reactor (spent fuel storage)

AR--at reactor (spent fuel storage)

BWR--boiling water reactor

CANDU--Canadian natural-uranium, heavy-water-moderated and -cooled power reactors

DOE--U.S. Department of Energy

DWK--Deutsche Gesellschaft fuer Wiederaufar Keitung von Kernbrennstoffen

EPRI--Electric Power Research Institute, Palo Alto, California

E-MAD--Engine-Maintenance Assembly and Disassembly hot cell facility, NTS

FRG--Federal Republic of Germany (West Germany)

GCR--gas-cooled reactor

GNS--Gesellschaft für Nuklear-Service, FRG

HFEF/N--Hot Fuel Examination Facility/North, INEL

HTGR--high-temperature gas reactor

ICPP--Idaho Chemical Processing Plant, INEL

IFSF--Irradiated Fuel Storage Facility, INEL

INEL--Idaho National Engineering Laboratory, Idaho Falls, Idaho

KWO--Kernkraftwerk Obrigheim GmbH, FRG

KWU--Kraftwerk Union, FRG

KWW--Kernkraftwerk Wuergassen, FRG

LMFBR--liquid metal fast breeder reactor

LWR--light-water reactor

Magnox--magnesium alloy cladding for gas reactor fuel; typical composition = Mg-0.8 Al-0.0025 Be

NRC--U.S. Nuclear Regulatory Commission

NTS--Nevada Test Site

PHWR--pressurized heavy-water reactor

PNL--Pacific Northwest Laboratory, Richland, Washington

PWR--pressurized water reactor

SCC--stress corrosion cracking

SS--stainless steel

TVA--Tennessee Valley Authority

WNRE--Whiteshell Nuclear Research Establishment, Pinewa, Manitoba, Canada

WR-1--Whiteshell Reactor 1; heterogeneous, uranium, organic-cooled

DISCUSSION

C.J. Ospina, Switzerland

It was mentioned in the paper, that the EIR dry-storage cask for DIORIT fuel will store fuel with aluminium cladding. This statement is incorrect. The fuel has Zr-2 cladding and is enclosed in aluminium cooling tubes.

A.B. Johnson, United States

Thanks for that correction. I was confused by the fact that aluminium is the material which is visible as the fuel is stored in the spent fuel pool.

G.A. Brown, United Kingdom

You have mentioned that much of the US fuel will be long-cooled in pools and may have a small number of incipient clad defects. It seems unlikely that it could be guaranteed that all moisture will be detected or removed. Do you think there could be a problem with moisture trapped in containers or casks which will eventually condense into liquid and is there a research programme needed in this area ?

A.B. Johnson, United States

We plan to conduct some evaluation of the moisture inventories in rods which developed defects in-reactor. For the long-stored fuel, the presence of water may be a minimal problem due to low storage temperatures but consideration of the consequences is justified.

M. Peehs, Federal Republic of Germany

You stated at the beginning of your presentation that irradiation annealing occurs already below 400°C. Annealing is a thermally activated process and should be completed in short time above 450°C. Do you believe that in the time the fuel stays in the 400°C region annealing occurs so far that hardening effects can be neglected calculating uniform elongation ?

A.B. Johnson, United States

For relatively short residence at 400°C, e.g. in casks where fuel temperatures are dropping rapidly, annealing may be minimal. My point is addressed more to the case of hot cell tests where the fuel may reside at 400°C for several months. My concern is not that the annealing jeopardizes the fuel cladding properties. In fact it tends to increase the ductility. The only point I make is that the mechanical condition of the cladding no longer has a clear relationship to that of unannealed cladding. Thus there must be some care in drawing conclusions about cladding performance extrapolation between annealed and unannealed cladding.

<u>M. Peehs</u>, Federal Republic of Germany

You showed during your presentation a BWR FA with a thick crud deposit on the surface. Crud layers are thicker in BWRs than in PWRs. Also the crud deposit shown is a-typical for BWR FA I have seen from KWU-reactors. Do you feel that crud may produce harmful effects on cladding integrity ?

<u>A.B. Johnson</u>, United States

I agree that BWR crud deposits are generally thicker than PWR crud deposits. Also, as I said in the talk, the two types of deposits differ in composition and morphology. The crud deposit colour suggests Fe_2O_3, which is typical of most BWR crud deposits. We of course cannot tell what the crud thickness is from the exterior appearance, so I cannot say whether or not the thickness is a-typical. I do not believe that crud will produce major impacts on dry storage, but the effects, particularly during fuel retrieval and in storage system decommissioning, should be anticipated.

<u>J. Fleisch</u>, Federal Republic of Germany

I would like to comment on the figure showing fuel cladding temperatures for different operating dry storage systems. This figure illustrates the maximum local fuel cladding temperature, which is e.g., in the case of the Würgassen cask demonstration, restricted to the 4 center fuel assemblies (out of 16) and in these elements, to the middle section of a few center fuel pins. Therefore the figure should be specified. Does an axial temperature gradient affect fuel cladding integrity ?

<u>A.B. Johnson</u>, United States

I appreciate your clarification regarding the Würgassen test temperatures. Hydrogen in zircalloy cladding tends to diffuse down a temperature gradient. However, there have been axial temperature gradients for cladding in-reactor without indications of major hydrogen re-distribution. I have mentioned possible annealing of radiation damage, but I don't see that as a problem.

DRY STORAGE CONCEPTS AND THEIR THERMODYNAMIC LAYOUT

R. Schönfeld
NUKEM GmbH
Hanau (FRG)

ABSTRACT

The two favourable dry storage concepts being under consideration in the Federal Republic of Germany are presented and the physical behaviour or natural convection cooling with air is explained. With the three examples cask store, vault storage horizontally and vertically arranged the main thermodynamical design parameters and their influence on the efficiency of the cooling system and on the temperature distribution inside the store and of the stored material are discussed. Moreover, the importance of the fulfilment and the harmony of all safety criteria and the difficulties while to do so are carried out especially with the vault store.

RESUME

Les deux conceptions favorables de dépôt pour le stockage sec étant étudiées actuellement en Allemagne fédérale sont représentées et leur comportement physique de refroidissement de convection naturelle sera expliqué. Pour trois types de dépôt, à savoir stockage de conteneur, stockage de type bloc arrangé horizontalment et verticalment, sont discutés les paramètres thermodynamiques principales ainsi que leur influence sur l'efficacité du système de refroidissement et sur la distribution de température à l'intérieur du dépôt et dans les matériaux stockés. En plus, l'importance de la réalisation et de la coincidence de tous les critères de sécurité et toutes les difficultés entrant en jeu seront mis en relief specialment pour le dépôt de stockage de type bloc.

1. Introduction

The safe interim storage of spent fuel elements from nuclear power plants is an important part of the back end of the nuclear fuel cycle. With progressing time and continously rising demand for interim storage capacities a suitable and satisfactory solution for this part of the nuclear fuel cycle gains more and more importance.

An essential aspect of interim storage is the fulfilment of the following safety-relevant criteria:

1. safe enclosure of radioactive materials
2. safe and simple handling
3. undercritical arrangement
4. radiation shielding
5. safe heat removal

The safe heat removal is especially important as this aspect has consequences on the other criteria in the above.

To avoid over-heating of the stored, heat generating material and the surrounding safety barriers a reliable removal of the generated heat must be provided. The safe heat removal in an interim store can be achieved by forced or natural convection using water or air. Principally due to safety reasons but also due to financial reasons natural cooling with air is being preferred at present.

The two variants being under consideration in Germany are the vault and the cask store.

2. Dry Storage Variants with Natural Convection

2.1 Cask Store

In a cask store (fig. 1) the spent fuel elements are contained in thick walled casks (fig. 2), stored in a hall and cooled by the ambient air.

The safe enclosure of the radioactive material and the radiation shielding is ensured by the corresponding gas tight and thick-walled cask. The criticality safety and the safe heat removal to the outer surface of the cask is ensured by the design of the cask basket. The outer surface is constructed in such a way that the heat generated within the cask can under certain conditions be sufficiently transferred to the ambience.

The safe heat removal from the storage casks to the ambience is achieved as in a vault store via natural convection of air.

2.2 Vault Store

In a vault store (fig. 3) the spent fuel elements are stored in storage racks which are arranged in shielded cells. These cells are connected directly or indirectly with the ambience via openings or ducts. In the case of direct connection with the ambience the stored materials are directly cooled by the ambient air, whereas in the case of indirect connection the cooling occurs via intermediate heat exchangers. In the case of direct cooling the safe enclosure of the radioactive materials is ensured primarily by a gas tight fuel element canister and secondarily by a storage building designed against external events such as plane crash, earthquake, etc. In the case of indirect cooling the heat exchanger presents a further barrier, thus it is possible to dispense of the canister.

The undercritical arrangement of the fuel elements is given by the correspondingly designed storage racks and the safe and simple handling of the radioactive materials by cranes and manipulators specified in norms and standards.

The radiation shielding is carried out in two steps. The direct radiation is attenuated by thick concrete walls as well as lead and steel shielding doors. The radiation scattering in the inlet and outlet ducts is reduced by internal structures and deflections to the permissible values.

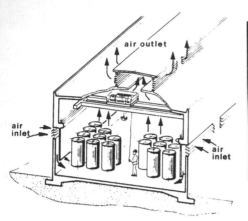

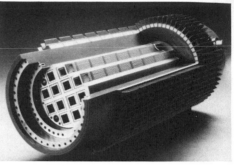

fig. 2: TN 1300 cask

fig. 1: cask store
(schematically)

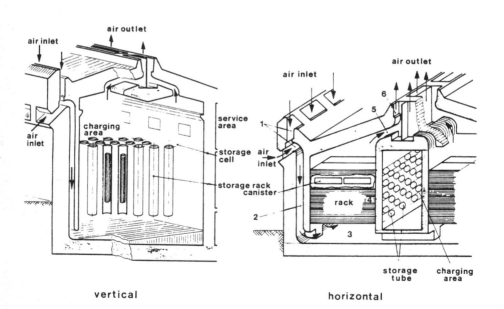

vertical   horizontal

fig.3: dry store
(schematically)

The safe heat removal from the stored material to the ambience is achieved as in the cask store by natural convection with air.

3. Concept of Heat Removal

Through the air inlet opening protected by bird gratings and the air inlet duct the ambient air flows to the heat producing stored material in the storage region where it is warmed up. The resulting change in density due to this warming produces buoyancy causing the warmed air to flow back through the air outlet ducts and the air outlet openings to the ambient.

The natural flow through the storage and the resulting cooling of the stored material is maintained by the continuous energy supply to the air within the cask region. This energy, the decay heat, is transformed to kinetic energy by buoyant forces. The natural convection is an equilibrium state resulting from the supply of internal and potential energy in the upward streaming air and the energy dissipation through friction and eddies in the ducts.

The heat transport from the stored material to the cooling air occurs in two steps. Through heat conduction the heat generated in the stored material is transported to the canister surface. From there it is transferred to the cooling air directly by convection and indirectly by heat radiation.

The essential advantages of this cooling system based on natural convection are:

- the inherent safety against the loss of coolant accident because due to the atmosphere an approximately infinite reservoir is present
- the system is independent of external energy sources
- the system is self-regulating, i.e. rising temperatures lead to an increase in the buoyancy which further leads to an increase in the mass flow of the cooling air resulting in improved cooling.

From the last point it follows that each canister sucks in exactly the quantity of air necessary to achieve its thermodynamical steady state.

4. Engineering Design

For the engineering planning and design of a dry store with natural convection cooling it should be noted that contrary to a water cooled store the cooling air negligibly attenuates the radiation. As a consequence the shielding has to be ensured by sufficient constructive measures.

The gamma und neutron shielding of the direct radiation is achieved as follows:

- in a cask store essentially by the thick-walled steel and iron cask fitted with a special neutron absorber. In addition, further shielding is attained by the ceiling and walls of the store building.
- in a vault by the thick concrete walls and ceilings as well as by the thick steel an lead shielding doors at penetrations and openings.

Especially in a vault special attention has to be paid to adequate attenuation of the radiation scattering in the air inlet and outlet ducts as here opposing demands arise from the shielding and thermodynamical point of view. For instance, straight ducts with large cross sections are desirable for an optimal cooling whereas ducts with small cross sections, built-in obstacles and large angular bends lead to an optimal attenuation of the radiation scattering; this can be clearly seen in fig. 5 and 6.

The harmonizing of these two points in respect to the fulfilment of the safety relevant criteria presents one of the essential problems

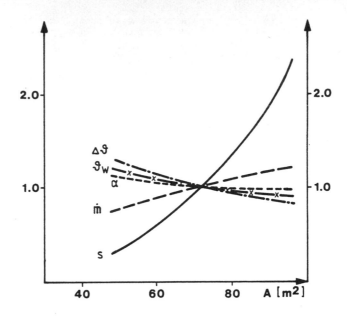

fig. 4: influence of inlet /outlet air
duct cross section A

(vault vertically arranged)

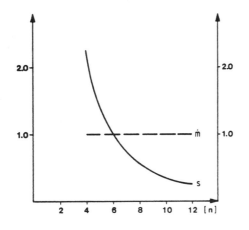

nomenclature :

"parameters normed to the
   design point „

$\Delta\vartheta$ temperature increase of
   cooling air

$\dot{m}$ mass flow of cooling air

α heat transfer coefficient

ϑ_W canister surface
   temperature

s attenuation coefficient

fig. 5: influence of number
of inlet / outlet
air ducts

(vault vertically arranged)

in the design of dry stores with natural convection. For solving
these problems various codes are used to deal with the thermodynamics and shielding; the majority of these codes were developed
at NUKEM.

5. Thermodynamic Design

5.1 Operating States

Basically one should differentiate between steady state and
transient operating states. Strictly speaking all operating states
and events in the store are transient due to the variable heat
generation of the radioactive material to be stored. They can be
classified as long-term and short-term events. Long-term events
cover a time of days, weeks or months. Such events are for example:

- increase of heat generation in the store due to charging with new
  canisters
- decrease of the heat generation in a completely filled store due
  to radioactive decay.

In the case of such events the dynamical behaviour of the cooling
mechanism can be neglected so that quasi-steady state calculations
can be performed. Short-term events are fast occurring events, such
as:

- temperature and pressure fluctuations in the ambience and conse-
  quentually the state of the entering ambient air
- charging with the first or initial set of canisters in the
  originally cold store.

During these events especially the dynamic behaviour of the cooling
mechanism is of special interest, which means that transient calculations must be performed.

The quasi-steady operating states and the resulting temperatures are
of special importance for the engineering design of a dry store,
whereas the transient events and the resulting temperatures are of
greater interest for the safety analysis of a dry store.

5.2 Mathematical Modelling

The thermodynamical events in the flowing coolant air are described
under consideration of assumptions and simplifications of the
balance equations for mass, momentum and energy. The store cooling
system is divided for physical reasons into three main regions:

- air inlet
- storage hall or storage cell
- air outlet.

For steady state and quasi-steady state considerations the modelling
is one-dimensional in the flow direction with the cooling air being
treated as an incompressible medium and described by the ideal gas
equation. In the inlet and outlet regions the flowing is assumed
isothermal. Within the storage hall or storage cell the heat is
transferred from the stored heat generating material to the cooling
air and causing an increase in cooling air temperature.

The heat transfer from the heated canister surfaces or fuel elements
to the coolant, which can be investigated numerically and analytic-
ally only with considerable expenditure, are described by experimentally proven transfer functions.

The heat transfer within the thick-walled storage cask is described
by means of finite difference models using Fourier's equation. The
heat transfer within the fuel rod bundle as well as from the actual
fuel element to the surrounding tube walls in casks or inner walls
of the storage canisters takes place by conduction, radiation and
convection. The model for the mathematical formulation is that a
single rod within the fuel rod bundle is surrounded by eight
neighbouring rods and in the wall region of the canister case or in
the wall region of the storage cask sees a defined wall section /9/.

The heat transfer between the various gas-filled spaces formed by the fuel rods is described by a convective model. Detailed descriptions of the different models are given in /1, 2, 4, 9, 10/.

5.3 Verification of the Mathematical Model

Experimental investigations of the thermodynamic relations of interim stores or scale models with natural convection have not been carried out or the data were not available so that a complete verification of the computational model was not possible. But a number of experimental investigations of single phenomena do exist. These results were used for a partial verification of the mathematical model.

In the framework of one of our experimental programmes investigations were carried out on the convective heat transfer from the surface of the stored material to the cooling air as well as on the flow of cooling air through the storage rack /2, 3/. The storage concept - cylindrical storage canisters in a vertical, cylindrical storage tube - was constructed as a scaled-down, experimental set-up. Using this pilot facility it was possible to demonstrate the technical feasibility of the vertical storage using natural convection. It was shown that the known relationships for the heat transfer coefficients for forced convection in concentrical annullis agreed well with the experimental data.

By means of a pilot facility the constructive requirements for a uniform flow and distribution of cooling air in the storage tubes was experimentally investigated. The characteristic parameter was found to be the ratio of the inlet chamber cross section to the sum of the annulli cross-sections. A uniform flow and distribution of the cooling air is achieved when the ratio exceeds a minimum value.

The mathematical model to evaluate the rod temperature of the fuel elements with quadratic rod arrangement was verified by means of experimental and empirical results /5, 9/ as well as using benchmark calculations with the HEATING 5 code /11/.

5.4 Codes

The characteristic thermodynamical parameters of interim stores using natural convection were calculated based on the previously described general mathematical models using various codes. The systems of coupled differential equations of the various regions are solved iteratively within the codes via the buoyancy relationship applying a Runge-Kutta procedure of the fourth order.

The results of the computational model and the codes based on these models were verified by the above mentioned experiments and published experimental data.

A survey of the codes available and applicable for thermodynamic design are presented in the following table.

Table 1: Codes used for thermodynamical layout of dry stores

Name of the Programme	Description	Ref.	Source
HEATING 5	A programme to solve multidimensional, steady state and/or transient heat conduction problems in Cartesian, cylindrical and spherical coordinates	/11/	Union Carbide Corporation
ZYLIND	A programme to calculate the steady state, one-dimensional temperature distribution in cylindrical multishell elements		NUKEM GmbH

(Table 1, cont.)

SHVELA SHHOLA	Programmes to calculate the steady state flow and temperature distribution in a vault with vertically/horizontally arranged heating elements	/1,2,4, 9/	NUKEM GmbH
SHTBL	A programme to calculate the steady state flow and temperature distribution in a cask store with natural convection air cooling	/1,10/	NUKEM GmbH
HWTTLB	A programme to calculate the steady state and transient temperature distribution of cylindrical casks		NUKEM GmbH
SHVBE	A programme to calculate the steady state temperature distribution of a fuel element with square- arranged fuel rods	/2,9/	NUKEM GmbH

6. Results

The main thermodynamic design parameters of an intermediate fuel element store cooled by natural convection with air are:

- the fuel element heat generation rate
- the fuel rod heat generation rate per length
- the geometry.

The last one is a very complex parameter because it contains a number of "subparameters", for example the geometry of

- the fuel element, the canister, the cask
- the air inlets and outlets
- the air ducts
- the storage cell
- the storage rack (height, storage tube, tube arrangement)
- the height of the store.

Experimental and theoretical studies have shown that the design of the air inlet and air outlet regions as well as the design of the storage rack or the cask arragement are the most important geometrical design criteria.

As you can see in figures 5 to 9 which are all normed to the design values, there are by all means different influences of the parameters with the certain storage concepts. In the case of the vault store the favourable flow design of the air inlets, air outlets and the air ducts is very important because the cooling conditions inside the store decrease with growing flow resistance (fig. 7). This influence is not so important with the cask store. Because of the nearly neglectable effects of radiation scattering there is more flexibility in the construction of the inlet and outlet regions.

Figures 8 and 9 show the influence of the storage rack geometry of a vertically and a horizontally arranged vault. You can see from the curves that this parameter is also an important one. It depends also on other safety criteria (criticality and mechanical integrity) but normally there are no difficulties with these aspects.

The point mentioned above is also valid for the cask store. The arrangement of the casks and the center line distance between the casks inside the storage hall is very important for the unrestrained flow of the cooling air to the surfaces of the casks. Especially this point has not been sufficiently examined at the moment, so that it is not possible to make quantitatively safe statements on the influence of this parameter.

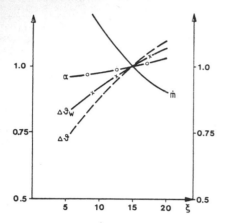

fig. 6: influence of the sum of inlet / outlet pressure drop factor

(vault vertically arranged)

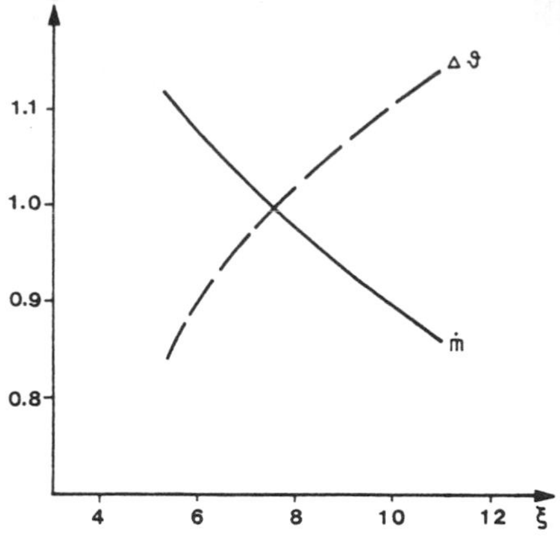

fig. 7: influence of the sum of inlet / outlet pressure drop factor

(cask store)

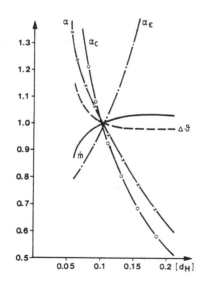

fig. 8: influence of the hydraulic channel diameter

(vault vertically arranged)

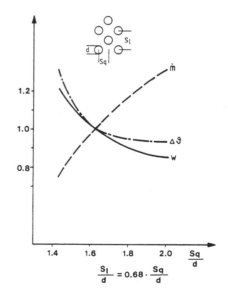

$$\frac{S_l}{d} = 0.68 \cdot \frac{S_q}{d}$$

fig. 9: influence of tube arrangement

(vault horizontally arranged)

Last but not least it can be pointed out, that the thermodynamic design of dry stores cooled with air by natural convection is based on extensive research work and first practical results. Furthermore, especially analytical and experimental work is necessary to analyse the several accident conditions and their time dependency. Also intensified work on the problem of radiation scattering in the air ducts should be carried out to reach an equivalent high research standard as in the field of thermodynamic design.

7. References

/1/ Schönfeld, R.; Hame, W.: "Physikalisches Modell zur Berechnung der stationären und instationären thermofluiddynamischen Größen in Luftlagern", NUKEM-401, 1978.

/2/ Hame, W.; Klein, D.; Pirk, H.: "Inhärent sichere Luftkühlung bei der Lagerung selbsterhitzender Radionuklidkonfigurationen", Kerntechnik 35 (1980) 2.

/3/ "HTR-Zwischenlager: Dokumentation experimentelle Untersuchungen", NUKEM 415 (12.1978).

/4/ Jeschar, R.; Tenhumberg, M.; Gardeik, H.O.; Schönfeld, R.; Klein, D.: "Bestimmung des stationären Temperaturniveaus im Zwischenlager für abgebrannte Brennelemente", BWK 31 (1980) Nr. 5

/5/ Mc Cann, R.A.: "Thermohydraulic Analysis of a Spent Fuel Assembly contained within a Canister", Pacific Northwest Laboratory, 1980.

/6/ Schönfeld, R.; Hame, W.: "Thermodynamics of Free Convection Air Cooled Storages", Proc. European Nuclear Conference 1979, pp. 553-555, ENS, Hamburg, 1979.

/7/ Hame, W.; Schönfeld, R.: "Sensitivitätsuntersuchungen zu im Naturzug mit Luft gekühlten Brennelementlagerkonzepten", Proc. Jahrestagung Kerntechnik 1980, pp. 113-116, KTG, Berlin, 1980.

/8/ Knappe, O.W.; Schönfeld, R.; Wingender, H.J.: "Dry Storage Systems with free Convection Air Cooling", Conference on Storage of Spent Fuel, IAEA, Las Vegas, 1980.

/9/ Schönfeld, R.: "Berechnungsmodell zur Untersuchung der Wärmeabfuhr aus einem naturzuggekühlten, horizontalen Zwischenlager für selbsterhitzende Materialien", will be published in 1982.

/10/ Dorst, H.J.; Schönfeld, R.: "Thermodynamik eines naturzuggekühlten Behälterlagers für radioaktive Materalien", will be published in BWK 33 (1982) Nr. 11.

/11/ NEA-DATA-BANK: (W. O. Turner, D. C. Elrod, I. I. Siman-tov): HEATING 5, An IBM 360 Heat Conduction Program, Union Carbide Corporation, Nuclear Division, Oak Ridge, ORNL/CSO/TM-15, (D. C. Elrod): REGPLOT, A Plotting Program to Graphically Check HEATING 5 Input Data, K/CSD/TM-12, (D. C. Elrod, W. O. Turner): HEATPLOT, A Temperature Distribution Plotting Program for HEATING 5, K/CSD/TM-11 1980 modified by NUKEM GmbH to the use on CDC-Cyber-Computers

DISCUSSION

I. Oyarzabal, Spain

For a cask concept, have you analyzed or tested natural convection inside the cask, that is between basket and walls ? If so, are you able to quantify the importance of natural convection as compared with radiation and conduction ?

R. Schönfeld, Federal Republic of Germany

We have analyzed it as a first step numerically, based on experimental results carried out in the United States.

The quantification of the convective part of the heat transfer depends on different parameters, for example :

- geometry;
- fill gas;
- temperature range.

In view of the small gaps between fuel element and surrounding storage tube, the small hydraulic channel diameters inside the assembly and the spacers, which are arranged nearly every half meter in a PWR-assembly, the geometrical conditions are not good for intensive convective heat exchange.

If there are no greater gaps in the basket where the convective flow can take place, convection can be ignored in the case of a helium filled cask.

In the case of air or nitrogen filled casks the convection part increases with decreasing temperature levels.

The basket we analyzed was a compact aluminium basket as used with the TN 1300 cask. This basket is primarily designed for conduction heat transfer.

At a temperature level of 400°C for the maximum fuel rod temperature and 4 kW fuel element heat generation rate, the convective part of the total heat transfer coefficient was found to have an influence of less than 10 % on the maximum fuel rod temperature. In the temperature range of 200°C-300°C the convection part is still higher and I think may reach values of up to 50 %.

H.J. Wervers, The Netherlands

Can you mention reliable values of radiative emission coefficients for zircalloy surfaces under PWR or BWR conditions ?

R. Schönfeld, Federal Republic of Germany

From several US-investigators and from the KWU I got the information that the emission coefficient is $\varepsilon \geq 0.8$.

Construction of JRR-3 Spent Fuel Dry Storage Facility

M. Adachi
Japan Atomic Energy Reseach Institute
Tokai, Japan

Abstract

To store the JRR-3 metallic natural uranium spent fuel elements, dry storage facility has been constructed in JAERI. This
facility has a capacity of about 30T of uranium.
The elements are placed in encapsulated canister, then stored in
drywell in the store.   The store is basically an ordinary concrete
box, about 12m long, 13m wide, and 5m deep.   The store comprises a
10 x 10 lattice array of the drywells.   The drywell consists of a
stainless steel liner which is 2.5m deep, 36cm ID and 0.8cm thickness. A drywell also has an air inlet, outlet pipe for radiation
monitoring and a shield plug in carbon steel for radiation protection.   A canister which consists of stainless steel with 0.5cm
thickness contains 36 elements. Sealing of the canister is accomplished by fusion welding.

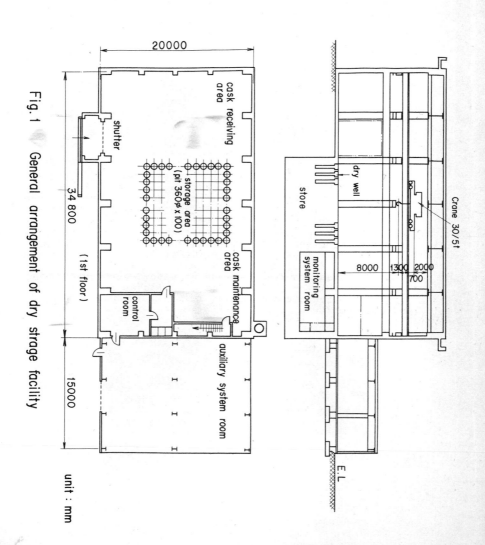

Fig.1 General arrangement of dry strage facility

unit : mm

1. Introduction

Japan Research Reactor NO.3 (JRR-3) had been a heavy water moderated/cooled tank type reactor of 10MW thermal power using metallic natural uranium fuels (aluminum cladding), which has been using a natural or 1.5% enriched uranium oxide fuels. It was prepared by Japanese staffs alone in order to experience the design, constructing and operating a reactor. The reactor has been operated for 20 years. Consequently, there have been 1800 metallic natural uranium spent fuel elements in JRR-3 spent fuel pond.
Originally it was expected that the elements would be removed to reprocessing plant. But it is now realized that the availability of reprocessing facility is unlikely to be adequate to accept the metallic natural uranium spent fuels.

To store these spent fuel elements, dry storage facility has been constructed at JAERI in February 1982. The basic arrangement of the facility is shown in Figure 1. This is the first away from reactor (AFR) spent fuel dry storage facility in Japan.

2. Spent fuel elements

Figure 2 shows JRR-3 metallic natural uranium fuel assembly and elements. A fuel assembly consists of shielding plug, cooling tube and three fuel elements. The elements are connected with pins. They are 1 inch diameter metallic uranium rods contained in aluminum alloy tubes in overall length about 950mm. The fuel assemblies are cut out to the elements in the pool. The specification of the element is the following:

Burn-ups	(max.)	800 MWD/T	
Cooling time	(min.)	2500 days	
Total fission product activities	(max.)	110 Ci	
Decay heat	(max.)	0.5 W	

For long cooling time, short-lived isotopes (e.g. I-131) have been decayed to negligible levels and dacay heat is very low.

3. Dry Storage Facility

The dry storage facility is located on the north area of JAERI Tokai Research Establishment. This facility has a capacity of about 30T of uranium. The facility consists of a storage building and an auxiliary systems room. The storage building contains cask receiving area, loading area, cask maintenance area and control room. The store is under the loading area. Spent fuel elements are contained in encapsulated canister, then stored in drywell of the store. Figure 3 gives the detail of a drywell. A canister which consists of stainless steel with 0.5cm thickness can contains 36 elements. Sealing of the canister is accomplished by fusion welding. The details of a canister are shown in Figure 4. Figure 5 shows the general arrangement of monitoring system. Slightly subatmospheric air environment in drywell is chosen for that any slight leakage will be inwards. The pressure in drywell is maintained at between 50 mb and 120 mb below the atmosphere. The environmental air in drywell is circulated by blowers during the working hours for radiation monitoring. At this period, radiation level of the environmental air is monitored by plastic scintillation counter. The location of leaky fuel elements and canisters can be determined by valve operation. The fuel elements don't need cooling because it's low burn-up and long cooling time. Accordingly, the cooling system is not provided. Pressure variation in this system is continuously monitored to find fission product gas (Kr-85) release. The system is also provided with dehumidifier so that humidity can be kept below 20% relative humidity to prevent corrosion of aluminum cladd-

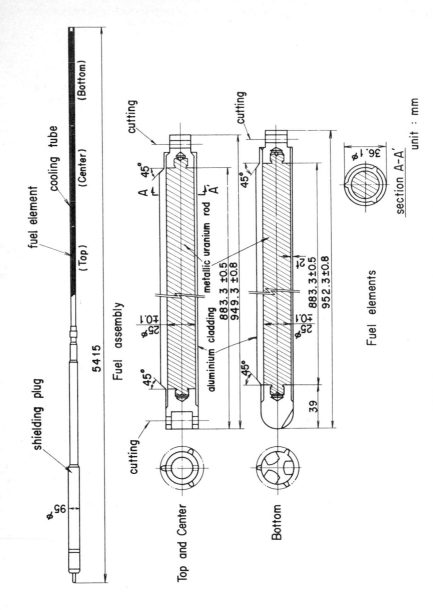

Fig. 2　JRR-3　metallic natural uranium fuel

unit : mm

ing. Usually this system has been held in closed circuit. For fil-
tration of air exhausted from system, fans and filters are provided.
The shield plug is bolted to the upper plane in drywell. The drywell interior is sealed by neoprene "O" ring packing between the
bottom of shield plug and the ledge in drywell.

4. Canister and Encapsulation

The canisters will be encapsulated at hot laboratory near JRR-3.
Hot laboratory has a number of hot cells that is constructed by
reinforced concrete shield walls with numerous lead glass viewing
windows. The cell is equipped with a number of remote handling
machines to support remote operations within the cell. Fusion welding of the canister is accomplished by TIG welding machine in the
cell shown in Figure 6. This machine is designed specifically for
remote operation. The canister, shown in Figure 4, consists of the
canister body and closure lid. The closure lid consists of a flat
disc and a handling head of stainless steel. This flat disc has
some lobes, which mate with dents machined into the canister upper
body. Following installation of the fuel elements into the canister,
the closure lid is fully threaded in. Canister is sealed by fusion
welding of a small lip of the closure lid to the upper surface of
the canister body. This small lip is machined as a part of the
closure lid. Following the seal welding, the evacuation/backfill
tube is connected to top of the handling head. The air in the canister is evacuated and the canister is backfilled with helium to a
pressure of the atmosphere. After helium filling is complete, the
evacuation/backfill tube is removed and a small plug is inserted
into the handling head. In this work, helium is maintained in the
canister by spring and metallic packing that is provided in the
closure lid. Then, sealing of the canister is accomplished by
fusion welding between a small lip of small plug and top of the
handling head. After the canister has been filled with helium,
various inspections are performed. These inspections include visual
inspections, helium leak check and surface contamination check.
Visual inspections are performed through the periscope and helium
leak check is performed by vacuum chamber in the cell.

5. Additional equipments

Dry storage facility is designed to accept dry fuel elements
in dry cask. From hot laboratory to dry storage facility, the
encapsulated elements are transported by the vehicle. The shipping
cask is stainless steel covered lead cylindrical structure, about
160 cm in nominal diameter and 260 cm high. It's total weight is
about 18 tons. The design of the cask and it's handling provisions
are such that an accidentally dropped cask will not release the
radioactive materials. The shipping cask is removed using the
overhead crane provided in the dry storage facility. The crane has
a capacity of 30 tons. Gamma ray shielding gate is used to position
and radiation protection when canister is loaded into the drywell.
This gamma ray shielding gate is box type and has a stainless steel
covered lead structure. If the leaky canister is detected at helium
leak check, the canister is cut by diamond cutter provided in the
decanning machine and the elements are taken off and recanned in
other canister. To evaluate the contamination, the manipulator
operator takes smear samples on the canister surface by filter
paper. If the canister is contaminated, the surface is washed in
the cleaning chamber which has a number of spray nozzles and equipment for water supply.

6. Safety evaluation

The facility is designed to meet the almost same stringent

safety requirements as applied to japanese nuclear reactors. The structure is designed to retain the integrity in general loading and transient operation, even in earthquake. The fuel elements are placed in the welded canisters whose design is such that an accidentally dropped canister will not release the radioactive materials by shock absorber. A shock absorber is honey-comb type and placed in the bottom of drywell. Even if a number of leaky canisters and fuel elements generate, the only isotope of importance which reaches the environment is Kr-85. But, the safety analysis shows that any release of Kr-85 is below the safety criteria limits.

The maximum temperature of the spent fuel elements will not exceed 45 ℃ during storage because the decay heat is removed by natural convection. Shielding wall has above 1.5m thickness, so radiation protection is adequate. Air inlet and outlet tubes are arranged with dogleg turns built in to prevent direct radiation streaming. Pre-test examination shows that the corrosion of aluminum cladding is negligible and obviously the elements do not need drying. The facility and handling provisions allow for retrieval and detailed inspections of selected canisters, including cutting the canister for non-destructive examination of the fuel elements at hot laboratory.

The results of the safety assessment indicate that the environmental impact of normal operation and possible accidents in dry storage facility is negligible small.

7. Conclusion

The dry storage facility is about 3 Km from JRR-3 and located on the North area of Tokai Establishment. We had expended a total of about 4 years from initial conception to completion of this facility. The dry environment was chosen because of that has significant advantages; it is safety, simpler, low operator dose rates, very reliable, negligible contamination and waste generation, and requires a significantly smaller operating staff when compared to wet storage. There has been no safety regulation for dry storage of spent fuels in Japan. Therefore, we have designed the facility to meet almost same stringent safety requirements as applied to japanese nuclear reactors. We believe that the dry storage facility will give us a high assurance that the spent fuel elements can remain in dry storage for very long time without deteriorating.

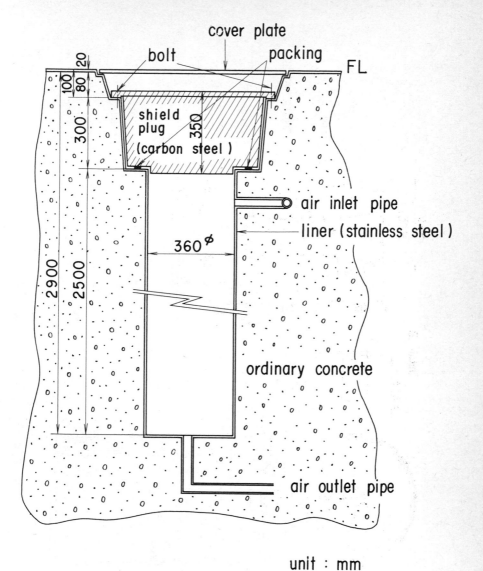

unit : mm

Fig. 3 Drywell of dry storage facility

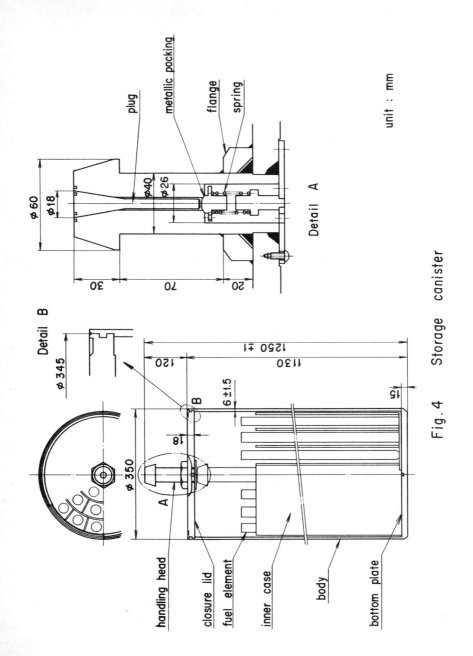

unit : mm

Fig. 4　Storage canister

plug
metallic packing
flange
spring

Detail A

Detail B

handling head
closure lid
fuel element
inner case
body
bottom plate

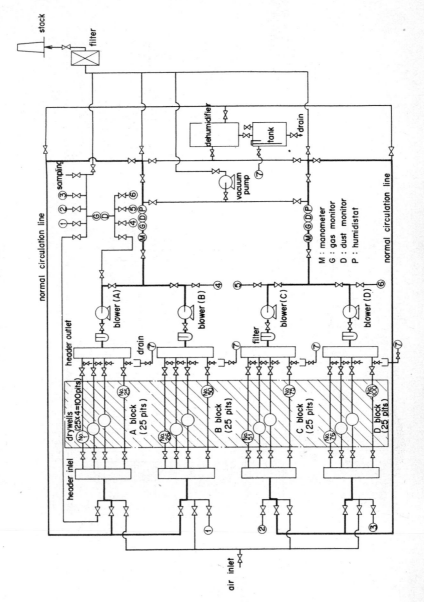

Fig. 5   Monitoring system of dry storage facility

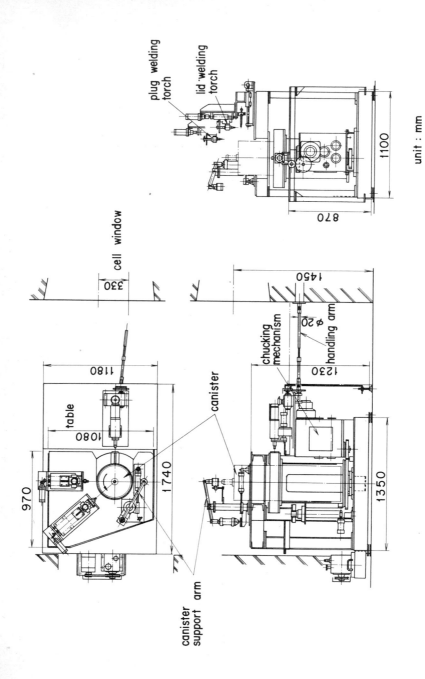

plug welding torch

lid·welding torch

unit : mm

1100

870

cell window

330

table

1180

1080

970

1740

canister

canister support arm

chucking mechanism

ø20

handling arm

1450

1230

1350

Fig. 6　Canister welding machine

- 170 -

DISCUSSION

M.S.T. Price, United Kingdom

Could you give us an idea of the capital cost of the JRR-3 spent fuel dry storage facility and comment on the capital cost. It would be interesting to know if you consider the facility to be over- or under-designed since there are no regulatory guidelines for dry storage in Japan.

M.A. Adachi, Japan

The total cost is about seven million dollars including the welding machine, cleaning machine, decanning machine, shipping cask and so on. I think it is too expensive. But at the present time storage is the most important problem for us because reconstruction of JRR-3 is being planned and we therefore have to remove these spent fuels to another facility.

R.J. Pearce, United Kingdom

If you have residual water present on a fuel element there is the possibility of uranium hydride formation if the element can is penetrated. Have you considered this ?

M.A. Adachi, Japan

Spent fuel is transported from JRR-3 to the hot laboratory by the inner case shown in Figure 4, and this inner case has many holes in the bottom plate. So, if it is lifted up from the pool, water in this inner case will be lost immediately.

G.A. Brown, United Kingdom

Many precautions are taken to ensure that storage conditions are very dry and therefore corrosion free. You use stainless steel extensively. Is this necessary ? We favour carbon steel because of low cost and easier welding.

M.A. Adachi, Japan

We don't know when the spent fuel will be reprocessed. This facility is constructed for spent fuel which cannot be reprocessed for a very long time.

Therefore, we used stainless steel extensively and I think it was a very reasonable selection in order to get the approval of the competent authority in our country.

H. Konvicka, Austria

You mentioned that there was no safety regulation available for dry storage in Japan. Has such a regulation been developed in the meantime ?

M.A. Adachi, Japan

We have the approval of the competent authority in our country for an ordinary technique to be used for Japanese reactors.

I have not heard of such a regulation being developed in our country.

INTERMEDIATE DRY STORAGE
OF THE SPENT FUEL OF REACTOR DIORIT

Dr. C. Ospina
Swiss Federal Institute for Reactor Research (EIR)
CH-5303 Würenlingen, Switzerland

ABSTRACT

The storage of the spent fuel of the reactor DIORIT is required, until it is either reprocessed or conditioned for final disposal. Different storage concepts have been studied as alternative to actual water pools; as a result it has been decided to store the spent fuel in a dry transport and storage cask.
    The project is well advanced, the cask is under construction and the fuel loading facilities are being implemented; this facilitates the decommissioning of the reactor DIORIT.

RESUMEN

El almacenaje de el combustible usado en el reactor DIORIT es necesario, hasta que se decida su reprocesamiento o acondicionamiento final. Diferentes conceptos para almacenar combustible han sido estudiados, decidiendose por un contenedor de transporte en seco.
    El projecto está bien avanzado, el contenedor se encuentra en construción y la infraestructura para el transvase de combustible se está implementando. Este projecto facilita el desmantelamiento de el reactor DIORIT.

Fig. 1
DIORIT FUEL ELEMENT

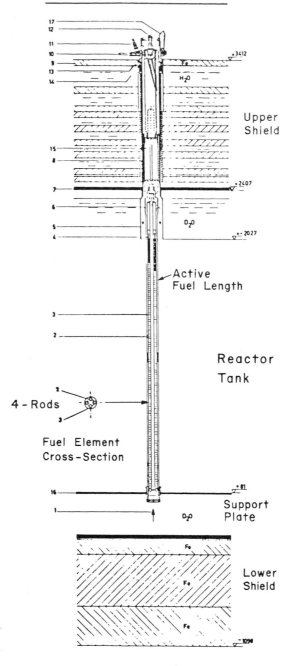

Upper Shield

Active Fuel Length

Reactor Tank

4 - Rods

Fuel Element Cross-Section

Support Plate

Lower Shield

1. Introduction

The reactor DIORIT was a heavy water research reactor operated at our institute. The reactor nominal power was 30 MW$_{th}$ and after serving during 17 years, it was shut-down in mid 1977 for final de-commissioning.

The fuel of the reactor was made so that four fuel rods were accomodated into a cooling tube (see fig. **1**) builing-up in this way a fuel element; of the total used fuel of 350 elements, about 40 % used natural uranium and the rest 2.2 % enriched uranium. The maximum burn-up reached was about 17.700 MWd/tu and the maximum rod power achieved was about 116 kw (see table **I**).

Basically the storage of spent fuel in water pools has been planned and used under the assumption that fuel would be reprocessed after only a short period of storage, and indeed few of the early DIORIT elements were processed abroad (Eurochemic and Marcoule). Subsequent developments in the late '70 have affected this policy and spent nuclear fuel now has to be stored for extended periods /⁻1⁻7 before its final route is been decided.

Since the shut-down of the reactor the discharged fuel (224 elements) and the fuel already awaiting for reprocessing (126 elements), have been stored in the reactor cooling pools.

2. Storage Alternative Systems

Two major reasons have moved us to search for an alternative storage scheme to water pools : first of all the reactor DIORIT is being gradually decommissioned and therefore the water pools are quite useful, e. g. to start the decontamination and conditioning of some reactor components for final disposal; secondly as the spent fuel might be stored for extended periods, say 10 or more years, there are no technical, economical or planning reasons to leave the fuel in the pools while they require active cooling systems and continous control (by the way, this reasoning holds for any nuclear power station and specially after its shut-down).

Therefore we have concentrated into alternative dry and passive storage concepts (no active systems been required), that facili-tates its transport anytime and that are economical while been advantagous for the decommissioning of the reactor (there is no more a reactor), and last but not least, that possess an inherent safety for long-term storage; this last is one of the best advantages to look for, if we regard the global picture of the LWR-plant Systems depicted in terms of component problems and materials-related causes (see table II).

We can see in the mentioned table above, that stress corrosion cracking (SCC), intergranular stress corrosion cracking (IGSCC) or just general corrosion are common primary causes for major pro-blems in LWR-systems either at high temperatures ($\sim$ 300° C) or low temperatures regimes (20 - 100° C) /⁻2, 3⁻7, working in water / steam media or just stagnant under water. Therefore the interest in having alternative intermediate storage systems, different from the actual cooling pools. Any other solution should allowed an indepen-dent storage of the spent fuel (away from reactor), while been inherently safe to permit a long-term and economically acceptable storage solution.

Obviously the only inherently safe storage alternative for spent fuel is dry passive-cooling; therefore our studies concentrated to-wards such concepts.

A first evaluation considered a dry-cooling vault concept, re-sulting it, in a relatively expensive, bulky and laborious design (requires fuel capsules) for such an small amount of spent fuel.

The next concept was the so-called concrete canister, as used today for reactor-Candu spent fuel. This concept proved to be econo-mical and technically amenable, nevertheless it required for our

Table I

DIORIT fuel inventory and activity

status : 10.08.77

Parameter		2.2% Enrich. Uranium	Natural Uranium	Total
Number of fuel element (FE)		196^x	140	350
U-235	g	17'620	4'187	21'807
U-total	g	1'445'411	1'039'368	2'484'779
Activity Ci	β	9'644'000	2'866'000	12'510'000
	γ	4'411'000	1'332'000	5'743'000
max. Burn up	MWd/tu	17'700	6'450	
Rod max. Power	kW	116.2	64.6	
Cool. Time	days	182	182	
Discharge Date		---	10.08.77 : 196^x FE	84 FE {20.11.73, 15.05.75 56 FE {14.09.76, 10.08.77	

x   plus 14 FE hardly irradiated

Table II

LWR SYSTEMS : COMPONENT PROBLEMS AND MATERIALS RELATED CAUSES

REACTOR COMPONENTS	WORKING MEDIA	KEY MATERIALS	MAJOR PROBLEMS	PRIMARY CAUSES
1. CORE	Water & Steam			
Fuel Rod (mainly BWR)	($\sim$300° C)	Zircaloy/ UO_2	-Pellet-Clad Interaction -Cladding Perforation -Wastage	-Stress Corrosion Cracking (SCC) -Corrosion, Hydriding
Fuel Assembly	Channel (BWR)	Zircaloy	Bowing and Dilation	-Creep/Corrosion
2. PRESSURE BOUNDARY	Steam & Water			
Vessel (mainly PWR)		Low Alloy Steel	Demonstrate 30 year Integrity in Presence of small Cracks	-Radiation Embrittlement and Corrosion Fatigue
Piping (BWR)	Steam & Water	304SS	Cracking in Weld Heat Affected Zone (HAZ)	-Intergranular SCC (IGSCC)
Piping (PWR)	Water	C Steel	Cracking	-Corrosion Fatigue
3. HEAT EXCHANGER	Steam & Water			
Steam Generator	(PWR)	C Steel/ Ni Alloy	Tube and Tube Support Distortion and Crack Possible Rupture of Tubes	-General Corrssion and SCC
Condenser		Cu-Ni Alloys	Tube Failures	-SCC, Erosion, and/or Westage
4. FUEL POOLS	Water (20-50° C)			
Fuel (LWR)		Zircaloy/ UO_2 Pool Piping (304SS) Pool components (304SS)	Pellet-Clad Inter. Cracking in Weld Heat Zone (HAZ) system impurities (CL^{-1}, Na ...) Crude	-IGSCC, Corrosion, Hydriding ind. galvanic, clad stress ?

Fig. 2

Dry Storage and Transport Cask
CASTOR I_c-DIORIT

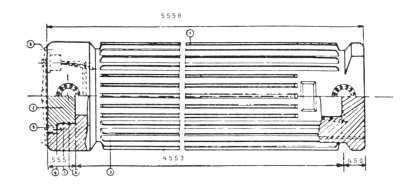

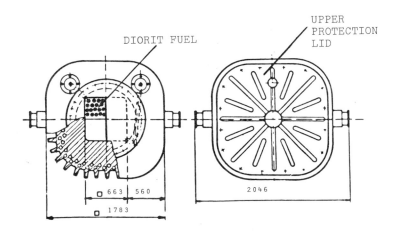

DIORIT FUEL

UPPER
PROTECTION
LID

specific fuel, a transport cask between cooling pools and the storage canisters, as well as laborious loading and un-loading operations; additionally the concrete canister is either designed nor licensed for transport further steps for the fuel management are required.

A most recent approach proved to be more adequate for our problem, namely the CASTOR-transport and storage cask /⁻5, 6⁻/ (as developed by GNS* in the FRG), that permits us the use of a dry passive-cooling concept, that simultaneously allows at-reactor and away-from-reactor storage, while being technically and economically sound.

Additionally and very important, there is no secondary waste production in this type of passive dry storage concept.

3. PRELIMINAR PLANNING

A detailed analysis of the mentioned CASTOR-cask showed that different versions of it are available. In order to choose the proper type of cask for the storage of DIORIT spent fuel, it was necessary to optimize the cask size and cost in order to minimize its total investment cost while maximizing the inner volumen utilization /⁻4⁻/. As parameters for the optimization we have used the following

from fuel side :

- length
- assembly grouping

from storage cask side :

- cask pay-load
- cask inner-volume
- cask height
- cask total cost

main constrain :

- minimum cask total weight

Once the optimum storage CASTOR I_c-cask (see fig. 2) was determined a tender was requested inclusive system-software, instrumentation, transport, etc.

The next step was to choose the appropriate loading-unloading alternative as a function of the optimum selected cask and the available installations at the reactor DIORIT.

After a detailed trade-off between reactor facilities and cask characteristics, a fuel loading variant (see fig. 3) under wet/dry conditions was found to be the best compromise, while minimizing the risk and total dose to any of the operators during the complete fuel transfer operations; a fuel transfer under water was out of discussion due to the place limitations for the storage cask into the cooling pools.

4. DEFINITIVE PROJECT

Extensive preparatory work has to be done before bringing the project into its final phase of execution, e. g. :

- cooling pool transfer equipment adaption and control
- reactor-hall equipment and facilities for the transfer operations has to be controlled for safety, radiation, protection, and redundancy under normal and emergency operational conditions.

---

* GNS : Gesellschaft für Nuklear-Service mbH, Essen, FRG

Fig. 3

Variant - III

Diorit spent fuel dry-storage
using a transport & storage cask-CASTOR

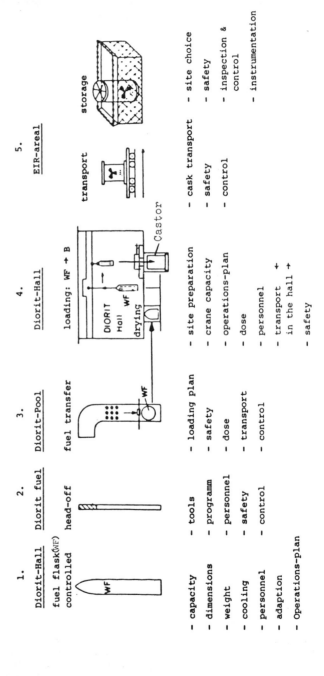

1. Diorit-Hall
fuel flask(WF)
controlled

2. Diorit fuel
head-off

3. Diorit-Pool
fuel transfer

4. Diorit-Hall
loading: WF → B

5. EIR-areal
transport   storage

- capacity
- dimensions
- weight
- cooling
- personnel
- adaption
- Operations-plan

- tools
- programm
- personnel
- safety
- control

- loading plan
- safety
- dose
- transport
- control

- site preparation
- crane capacity
- operations-plan
- dose
- personnel
- transport ↓
  in the hall ↑
- safety

- cask transport
- safety
- control

- site choice
- safety
- inspection &
  control
- instrumentation

- storage site choice, planning and engineering
- preparation of safety report for storage cask and storage site.
- integration of the preliminar's Institute planning with the storage cask manufacturer's engineering and planning group, in order to work-out the final project (timing, software, transport, etc.)
- planning of manpower required for the different project stages
- planning of the project financial requirements
- preparatory project-discussions with nuclear safety authorities in order to establish the requirements for its licensing etc.
- operations working-programm inclusive data adquisition and interpretation

After a year of work we were prepared for an order agreement for the CASTOR I_c-cask, the loading facilities have been planned and implemented, the final safety analysis report has been submitted and a plan for the final fuel loading activities has been coordinated (see fig. 4).

5. SAFETY ANALYSIS STUDIES, LICENSING AND REGULATORY

The Safety Analysis Report for the CASTOR I_c-cask has to be accomplished in order to satisfy the licensing authorities (ASK) for nuclear installations in Switzerland under the Atomic Federal Law regulations (Bundesgesetz über die friedliche Verwendung der Atomenergie); simultaneously the licencing and regulatory approval of the cask, by the safety authorities (ASK), is supported by the corresponding Type B(U)-license and transport certificate granted by the specific authorities of the country of origin of the cask (in this case FRG).

The Safety Analysis Report was prepared on a standard basis and properly adapted in order to reflect the specific characteristics of the DIORIT-spent-fuel; main topics of this report are :

- storage site
- cask systems
- thermodynamic and criticality analysis
- cask production programmes
- fuel analysis
- radiation and control analysis
- external and internal safety of cask

We are in the process of obtaining the mentioned permissions for validation in Switzerland by the corresponding licensing authorities (ASK).

6. PROJECT-STATUS AND CONCLUSIONS

The dry storage and transport cask has been ordered, its constructions has started and the fuel loading facilities are being implemented.

A fuel loading concept on wet/dry basis has been developed and an analysis of the storage site choices have been done, including an economic trade-off of factors affecting this decision.

It is expected to start the fuel loading operations by the end of 1982, followed by the implementation of a measurements campaign on the cask after loading it.

This project has been very valuable in assessing a new technology and the different problems encountered during its conception upto its execution and in this way facilitating the way for the final decommissioning of the research reactor DIORIT.

Finally, the value of this full industrial-scale project is large; its sound assessment and execution paves the way for at-reactor or away-from-reactor long-term intermediate large-storage projects. Conclusion : one cannot create experience, one must undergo it.

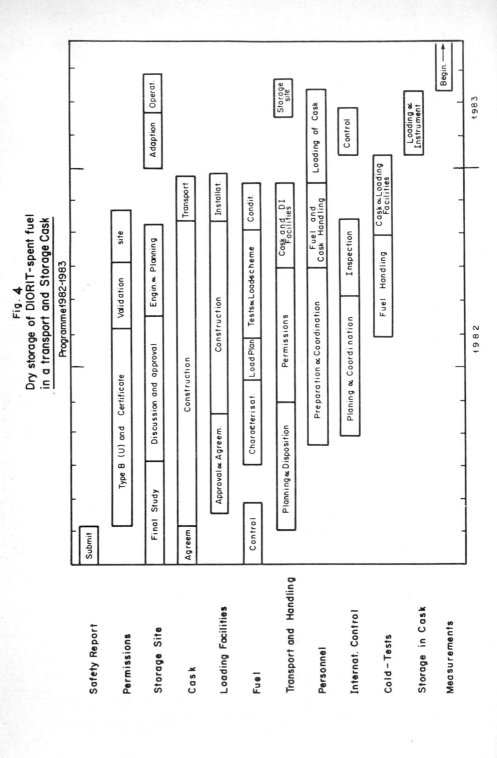

Fig. 4
Dry storage of DIORIT-spent fuel
in a transport and Storage Cask

Programme 1982-1983

Safety Report
Permissions
Storage Site
Cask
Loading Facilities
Fuel
Transport and Handling
Personnel
Internat. Control
Cold-Tests
Storage in Cask
Measurements

1982   1983

- 182 -

7. REFERENCES

_/_1_/_ Ospina, C. : " Trocken-Zwischenlagerung von abgebranntem Kernbrennstoff ", PrE-64, EIR-Switzerland, 06.08.1979

_/_2_/_ Divine, J.R., et al. : " Overview of spent fuel pool materials studies at Pacific Northwest Laboratory ", Int. Corrosion Forum; Toronto , Canada, April 06-10, 1981

_/_3_/_ Johnson Jr., A.B., et al. : " Assessment of stress corrosion cracking in a spent fuel pool pipe ", Int. Corrosion Forum; Toronto, Canada, April 06-10, 1981

_/_4_/_ Ospina, C. : " Technical overview of intermediate dry storage of spent fuel ", Proc. IAEA Advisory Group/specialists Meeting Las Vegas, Nevada, USA; Nov. 17-21, 1980

_/_5_/_ Baatz, H., et al. : " Cask Storage status and potential ", Proc. IAEA Advisory Group/specialist Meeting Las Vegas, USA, Nov. 17-21, 1980

_/_6_/_ Dyck, H.P. : " Konzeption von Zwischenlagern für abgebrannte Brennelemente ", Tagungsbericht, Entsorgung von Kernkraftwerken, Köln, Nov. 1980

Figure 5   Casting of CASTOR I_c for DIORIT Spent Fuel early 1982

DISCUSSION

H. Konvicka, Austria

Is there a cost estimate available for your undertaking, based on US $/kg heavy metal ?

C.J. Ospina, Switzerland

The current estimate available is about 120 US $/kg which should slightly increase after including the loading facilities cost.

R.J. Pearce, United Kingdom

Were the drop tests on the Castor flask carried out with fuel elements inside and, if so, have you examined the fuel elements ?

C.J. Ospina, Switzerland

The cask has been tested in accordance with the IAEA transport regulations for a type B(U) package, which includes among others, a free drop test from 9 m height onto a solid foundation; simulated but otherwise original fuel assemblies were used and inspected for mechanical compliance and integrity.

E.R. Johnson, United States

What was the capital cost of the air cushion device ? Can you briefly describe the device ?

C.J. Ospina, Switzerland

The cost of an air cushion device is about 8 000 US $/per module; we just lease them for the project.

The air cushion device consists of a module of 1.22 m^2 supporting surface and 0.70 m height; its weight is about 80 kg and it has a nominal load capacity of about 36 t. Each cask requires four such cushions.

H.J. Wervers, The Netherlands

Will the cask be housed in a storage building or will it be placed in the open air ? If so what are the arguments in favour of a building ?

C.J. Ospina, Switzerland

The cask will be housed in a storage building; in principle it could be stored in open air, but in order to facilitate the handling and inspection of the cask it is appropriate to house it in a weather protected building.

J.R. Haddon, United Kingdom

Please will you describe the system for transfer of fuel from the pool to the CASTOR cask at the Diorit reactor.

C.J. Ospina, Switzerland

The fuel is loaded under water into a transfer flask which is then transferred for dry-dry loading into the cask with the help of a loading platform (see Figure 3 of our paper).

N. Bradley, United Kingdom

My questions are directed to many authors who have mentioned the maximum design fuel temperature during storage. Would they comment on the two following provocative statements ?

(a) that for technical and economic performance reasons the cask proposals require 400°C; that requires an inert gas; that in turn means double containment with a frequency of test which brings the probability of more than one containment failing; that such cask designs are insensitive to whether defected fuel is present;

(b) that for single container designs with air atmosphere the temperature limit is due to oxidation of defected pins which must not release activity at a greater rate than that which permits detection, location and corrective action, e.g. below 250°C; that the build-up oxidation release in the container prior to a container failure, limits the temperature to less than 200°C with present oxidation data; that alternatively pre-failure oxidation could be prevented by inert gas-fill which does not require monitoring since container failure initiates detectable release based upon the rate of oxidation, e.g. 250°C is acceptable.

A.B. Johnson, United States

There is a wide range of residual heats for the current water reactor fuel inventory. Some fuel, even if consolidated, would not reach 400°C or even 300°C. Therefore a limit of 400°C is not required for much of the inventory. Where 400°C is needed as a limit, inert gas is required. Some level of monitoring appears to be prudent at least until the fuel temperature decreases below levels where oxidizing reactions are significant. If oxidants are absent, I am not aware of a mechanism which would cause significant degradation of a reactor-induced defect.

There is not yet sufficient oxidation data to establish a temperature limit for oxidizing atmospheres, though relevant studies are under way. Inert atmosphere operation is an alternative, but as indicated above is likely to require some level of monitoring.

M. Peehs, Federal Republic of Germany

Due to licensing requirements in the Federal Republic of Germany we need a multi-lid design with different barriers between the storage volume and the outside. Since we also have a pressurized control volume which is controlled during the storage, we have the possibility of having an inert gas atmosphere. Reviewing the different possible potential degradation mechanisms discussed in my paper, we realized that 400°C is a temperature - or more precisely an insertion temperature - which is safe and should not produce systematic cladding failures. If we conservatively assume having single failures, the inert gas atmosphere prevents a failure propagation or enhanced fission product release as discussed by our US colleagues, because they assume air in-leakage in single lid systems.

C.J. Ospina, Switzerland

(a)   The initial or insertion temperature of LWR fuel with a maximum canning temperature of 400°C, is a boundary to establish how old the fuel should be before dry storage (heat load and fuel capacity), as well as because of the limits in the internal pressure of fuel and storage atmosphere to prevent any failure of the fuel due to over-stressing.

Inert containment, double barriers and frequent tests are the means to guarantee the cask behaviour.

(b)   Single barrier cask designs are certainly less reliable and if air is the internal atmosphere, the risk of oxidation is much greater than using noble gases. A temperature reduction by insertion, say down to 250°C, alleviates the heat transfer and the mechanical design. Nevertheless, because of limiting extreme conditions (e.g. fire tests etc.), the cask design might require higher design boundaries.

M.S.T. Price, United Kingdom

The use of a cast flask implies a satisfactory theoretical model for mechanical behaviour particularly under impact conditions. By this I do not mean compliance with the IAEA Regulations (e.g. 9 m drop test) or compliance with a QA (Quality Assurance) programme. There is an ASME (American Society of Mechanical Engineers) Code Committee working on a code for forged flasks. Is there an equivalent initiative for cast flasks.

C.J. Ospina, Switzerland

The cask has been designed and built to meet the IAEA Transport Regulations and the Federal Republic of Germany (BAM and PTB) inspection and acceptance controls for a type B (U) package; this means that the package is unilaterally (U) approved and therefore, in accordance with IAEA Regulations, can be approved elsewhere.

P. Erlenwein, Federal Republic of Germany

You have shown the testing arrangement for the Castor cask to test the resistance against aircraft crash. The question : is the equipment for loading, handling and transporting the casks designed in a similar way, and if not, what is the safety philosophy behind this fact ?

C.J. Ospina, Switzerland

The design of the equipment for the loading and handling operations, is based on the quality and control requirements in nuclear power station installations under the safety rules that were granted for its safe operation; that is to say, the work is carried out in a controlled zone, whose safety is recognized and the equipment is compatible with the actual installation.

SESSION 4

Chairman - Président

A.B. JOHNSON
(United States)

SEANCE 4

OPERATING EXPERIENCE OF VAULT TYPE DRY STORAGE
AND ITS RELEVANCE TO FUTURE STORAGE NEEDS

E O Maxwell
Central Electricity Generating Board
Europa House, Cheadle Heath, Stockport, England

D Deacon
GEC Energy Systems Limited
Whetstone, Leicester, England

An outline description of the early passive cooled vault type dry stores for irradiated magnox fuel at the Wylfa Nuclear Power Station together with the valuable operating experience gained over many years.

An outline description of the world's first air-cooled vault type dry store (350 Te) and comments on its construction and successful operation. A description of the basic principles that were used in the design of this store and how these principles have been developed for use on vault type storage systems for oxide fuel and vitrified waste.

An examination of the basic parameters that the authors' consider should be used to measure the adequacy of the many storage options currently being considered around the world is included in order that a better assessment of the various systems may be obtained.

1. Introduction

For many years irradiated fuel has almost universally been
stored under water prior to it being reprocessed. Water provides a ready heat sink for the decay heat removal and also
provides shielding against the highly active fuel. The original requirements for storage were short term, the fuel being
stored until the decay heat output and radiation levels had
decayed sufficiently to allow transportation in suitable
flasks to the reprocessing facility. (Decay periods in the
order of 90-150 days).

There is now an increasing requirement to store fuel for very
much longer periods. Periods up to 100 years are being considered. It cannot be denied that to date there is far
greater experience in the pond system of irradiated fuel storage than any other form of storage. However, it is necessary
to take into account the fact that the early design of pond
installations took place before the design parameters for
irradiated fuel storage facilities became so exacting. It is
not suggested that pond type installations cannot be built to
the necessary safety standards but rather that when considered
against the latest design specifications other forms of
storage are likely to give better overall optimisation of
requirements.

In addition to the considerable experience in the design and
operation of dry storage in the UK it is also necessary to
highlight the fact that fuel has been handled 'dry' almost
without exception since the advent of nuclear power in Great
Britain. This tremendous wealth of design, manufacture and
operating experience of dry fuel handling machinery has enabled
us to confidently produce generic designs of dry fuel handling
machines for handling irradiated light water reactor fuel
between reactor storage pools and the dry storage facilities.
This is very relevant, no matter whether these be vault or
cask type stores located at the reactor or away from the
reactor.(Figures 1, 2,3 and 4)

2. Wylfa CO_2 Dry Store Cells, Designated Cells 1, 2 and 3

The concept of dry storage of irradiated fuel is not new. The
Wylfa magnox power station situated in North Wales was designed
with three dry storage modules when construction commenced some
18 years ago. These dry stores each have a capacity of 83
tonnes and they are cooled by completely passive means. The
irradiated fuel elements discharged on load from a reactor by
a Fuelling Machine are stacked 12 high in tubes connected at
the upper ends to a vessel containing the loading chute. This
chute distributes the fuel from the central access standpipe to
the tubes. The atmosphere within the tubes is CO_2 at a nominal
pressure of 3 psig and a purity of at least 99.8%. The primary
mode of decay heat removal from the fuel is by radiation to the
tube wall, conduction through the tube and then by natural
thermosyphon air cooling. (Figures 5 and 6)

A CO_2 atmosphere was chosen, as the limiting temperature for
both uranium and magnox in CO_2 is about 600°C, whereas in air,
the limiting temperature for uranium oxidation is 250°C. Very
low concentrations of air, eg, 0.2% could result in the oxidation of exposed uranium, if it were present, at a rate
approaching that of 100% air. Thus, a very high CO_2 purity is
required in order to allow the cells to accept fuel direct
from the reactors.

The operation and performance of these cells has been very satisfactory since they first came into operational service some 10 years ago.

3. Operational Experience with Cells 1, 2 and 3 (Figure 7)

3.1 The completely passive cooling regime has proved to be completely reliable.

3.2 The manning requirement has been limited to the personnel necessary to carry out the CO_2 purity checks. Frequency of checks, approximately once per week.

3.3 During operations to date many various heat load conditions from part load to full load have been experienced and the satisfactory performance of the store has been demonstrated. No detectable effects in store performance due to wind conditions have been found although the winds can gust at up to 130 KPH in this very exposed location.

3.4 The materials of construction of the store have been demonstrated to have an adequate life, even though these cells have their natural convection air inlet within 100 metres of the sea.

3.5 The Cells 1, 2 and 3 do not contribute any significant amount to the total station man-Sv commitment.

3.6 During testing and operation of these stores at Wylfa it has been demonstrated that the maximum measured fuel element temperatures are slightly lower than those predicted theoretically during the design process.

3.7 Post Irradiation Examination (PIE) of fuel having a dwell time of up to 4 years has shown it to be in pristine condition and indeed licencing approval has been obtained to allow this fuel to be reloaded into the reactor if necessary for further irradiation.

3.8 However, Cells 1, 2 and 3 had one drawback, ie, capacity. Generally, the CEGB's magnox station with cooling ponds have an average storage capacity of about 1 Te/MW of station output whereas at Wylfa it was about 0.3 Te/MW. Also Wylfa's throughput of irradiated fuel is about one quarter of the total arisings of the CEGB's magnox stations. Experience showed this capacity to be inadequate in the event of limitations in flask traffic or throughput at BNFL's Sellafield reprocessing plant.

4. The World's First Commercial Air Filled Vault Dry Store

Because of the above limitation, and Wylfa being the only stations where fuel is handled and stored dry prior to loading into transport flasks, it was decided in 1976 to increase on site storage of irradiated fuel. Bearing in mind that the existing CO_2 cells can accept freshly irradiated fuel direct from the reactors, these cells could be used as a decay store such that an air environment could be used for additional storage without exceeding fuel limiting temperatures, as discussed below:

4.1 Design and Safety Criteria (Figure 8)

Air was chosen as the environment for the fuel in order to eliminate the requirement for a store liner to seal to absolute standards. Although the construction of the liner is to high standards, and it is demonstrated to be leak tight on completion, it is very difficult to demonstrate that over very long periods the structure remains absolutely leak tight. The atmosphere within the store is maintained at a slight depression by means of an exhauster system discharging to atmosphere via a comprehensive filter system provided with generous redundancy. Should a slight leak develop in the store envelope it is readily recognised by a reduction in the store depression and remedial action can be taken. Obviously any leakage that does occur will always be into the cell and not contamination outwards.

The uranium ignition limiting temperatures of $250^{\circ}C$ for saturated air and $300^{\circ}C$ for dry air were set after consideration of existing experimental data. This data demonstrates that for irradiated uranium, the oxidation rate for the zone of highest irradiated swollen uranium at $250^{\circ}C$, is well below the characteristic of delayed ignition, and to date, no delayed ignition of irradiated fuel sections are known below these temperatures. In the absence of exposed uranium, the temperature limit would be set by the ignition of magnox, ie, $610^{\circ}C$ in saturated air and $625^{\circ}C$ in dry air.

Also, to inhibit the formation of uranium hydride which might cause spontaneous ignition in the event of fuel damage during handling, it is necessary to prevent condensation within the storage cell. This is achieved by means of a dehumidification plant.

Thus the following are the fuel design criteria for the air filled store:-

a)   the maximum fuel element temperature under normal operating conditions will not exceed $150^{\circ}C$

b)   the maximum fuel element temperature under the worst credible fault conditions will not exceed $200^{\circ}C$

c)   there is no condensation under the lowest credible temperature conditions ($25^{\circ}C$)

d)   during normal operating conditions the moisture level will be less than 30,000 vppm water or 50% relative humidity, whichever is the lesser.

Thus the margins before uranium ignition could occur is at least $50^{\circ}C$ and for magnox ignition it is at least $400^{\circ}C$. There are no known magnox ignition incidents as a result of uranium ignition.

4.2   Design Principles

It was decided to build two storage cells each with a capacity of 350 Te uranium. As it would be impossible to guarantee that the cell storage chamber could remain absolutely sealed throughout its life the atmosphere of the store should be subatmospheric to prevent outleakage of any contamination as previously described.

Natural convection was chosen for cooling the fuel within the store. An inherent advantage of natural convection cooling is that the cooling air does not need to be guided to the hot fuel. By careful design, it can be arranged that the fuel draws cooling air according to its needs.

The higher the heat output from the fuel, the greater
the buoyancy head and hence the greater the cooling
flow. The bulk heated air is then recirculated and its
heat removed by a conventional fan and heat exchanger
system embodying adequate redundancy. Five such fan/
heat exchanger units are provided of which four are used
for normal operation. When freshly irradiated fuel is
removed from the reactor (decay heat typically in the
order of 1 KW) it is stored in the existing CO_2 cooled
dry store cells for at least 150 days so that the decay
heat can reduce to a level acceptable to the air cooled
dry store. The heat output from a peak rated fuel element in the reactor, having decayed for 150 days is
approximately 46 watts including a 20% margin. The
gamma radiation to thermal output ratio was thoroughly
investigated for fuels under various burn-up and decay
conditions. Knowing this ratio for maximum burn-up, it
was possible to arrive at a value for gamma output
corresponding to the maximum permissible value for a
thermal output of 46 watts. Making allowances for drift,
and calibration of instrumentation, the maximum rating
of an element that can enter the store is less than
60 watts. Thus, as it is desired to set a limit on the
maximum possible thermal rating of fuel elements for
storage, it is only necessary to monitor the gamma
activity of the elements. This is achieved by
equipping the transfer machine (which transfers fuel
from the CO_2 stores to the air stores) with 3 gamma
detectors, to satisfy a high level interlock requirement,
that monitor the gamma dose rate corresponding to the
heat output of the fuel. The gamma dose rate chosen
allows the detectors to be set to reject fuel which has
a higher output than the predetermined limit. The
allowable heat rating of the fuel in the store and thus
the temperature limitations are in this way satisfied.

The equipment within the store is designed to eliminate
the need for maintenance. Such components that do
require maintenance are provided with maintenance
equipment designed to minimise radiological hazards in
the maintenance areas.

The irradiated fuel is adequately shielded during the
transfer operations from the CO_2 to the air cells and
in the air cells by the required thickness of reinforced
concrete.

Equipment and components are designed to prevent fuel
element damage at all stages of handling and storage.
The design is assessed for hazards, the severity of the
hazard determines the degree of protection required, ie,
integrity of interlocking. Any plant capable of shear-
ing a fuel element, is, in addition, fitted with torque
limiting devices and provision made to minimise the
damage in the event of a fuel element being dropped.
All electrical and cooling water supplies are duplicated
and on segregated routes.

5.   Layout of Plant

5.1   Limitation of Physical Parameters for the Wylfa Store

An important aspect of the construction of the air
cooled fuel stores at Wylfa was that the design and
construction had to be closely integrated with the
existing power station. The design and construction of

the new stores was successfully carried out without any interruption to the operation of the existing power station although they are closely associated structurally and operationally with the existing plant.

The requirement to ensure this compatibility with the existing structures has led to an embodiment of the principles outlined above which is less than optimum in some aspects of the store itself.

The Wylfa reactors are the largest magnox reactors that have been constructed (2 x 590 mw). To move the fuelling machines (150 Te) around the charge face by means of a conventional crane arrangement involving such a large span would have required significantly more expensive structural requirements than a remotely operated transporter system, and so this is the system that was adopted when the station was originally conceived. This transporter system gave a particular problem when it became necessary to choose the site for the new store (Cell 4) and also it had a significant effect on the resultant envelope for Cell 5. The working area for the fuelling machinery and transporters is shown diagramatically in Slide 15. During initial construction of the station, the working area was suitably reinforced to enable it to accommodate these loads. It can therefore be seen that it would be extremely difficult for the transporter/fuelling machine to move to the edge of the building without significant stiffening of the working floor over the chosen route. Civil work/structural modifications to stiffen the floor on an operational reactor was highly undesirable as it was the intention to construct the store whilst the reactors continued in operation. The only option available without structural alterations to the reactor building was to build a bridge across the existing access shaft. One end of the bridge could be supported on the area of pile cap already able to take such loads, and the other end of the bridge could be supported on an extension to the proposed new store, suitably designed to take these loads. In this way, the transporter and store transfer machine could run directly onto the top of the new store (Cell 4).

A fundamental disadvantage with this location was the fact that the original flask handling facility was already located in this position and the new store would therefore have to be located above this facility. The flask handling facility had to remain operational throughout the construction of the store which added to the problem.

Another disadvantage with this location was the fact that the base of the store would have to be located some 13.6 m above ground level and this precluded multi-element stacking within the store. A diagram showing the arrangement of Cell 4 is given in Figure 10.

5.2   Outline Description of Wylfa Cells 4 and 5 (Figures 9 & 10)

The dry store is a concrete box 60 m long, 11 m wide and 4.5 m high with walls that are 2 m thick. Six rows of twenty-five skips are positioned across the store. Each row of skips is supported on two drive chains which are in turn located on rails supported on concrete plinths

running the length of the store. The drive chains pene-
trate the walls at each end of the store. The chain
driving system is located outside the store with appropriate tensioning devices. A traverse mechanism for
moving skips from one row to another is located inside
the store at each end of the rows of skips.

Each skip comprises a matrix of blind tubes in a 12 x 16
array. The elements radiate heat to the natural convection cooled walls of the tube. The bulk temperature of
the area within the store is maintained at a satisfactory
temperature by 4 of the 5 heat exchanger units that
reject up to 750 KW of heat to the reactor ancillaries
cooling water system.

5.3   Construction and Commissioning (Figure 11)

Work commenced on site in September 1976. The initial
work was to identify and re-route various existing underground services to the station, eg, cables, radioactive
drains, fire mains, etc, in preparation for foundation
piling. In parallel with this operation, the civil and
plant detail designs were commenced and long lead items
of equipment were ordered for manufacture. Throughout
the design, construction and manufacturing phases the
civil structure and items of plant had to be approved by
an independent design assessor, a mandatory requirement
of the CEGB's Plant Modification Procedure. Independent
inspection of the plant after installation and prior to
commissioning was also a requirement. Before commissioning
of the plant could commence all of the relevant
commissioning documentation had to be approved in a
similar manner.

Commissioning of the plant was basically divided into
three sections:

1)   Transfer machine
2)   In-cell equipment
3)   Cell cooling/conditioning equipment

Each section was commissioned independently of the
other followed by final integrated system commissioning
tests. Before active fuel could be loaded into the store
the fuel safety document and commissioning test results
had to be approved by the independent design assessors
and the Nuclear Installation Inspectorate of the Health
and Safety Executive.

6.   Operational Experience (Figure 12)

The first of the two air cooled stores was commissioned for
operational use in September 1979 and was filled to capacity
(28,992 elements) by September 1981. The time taken to fill
the store was dictated by operational requirements of the
fuelling complex, ie, simultaneous operation of on-load
refuelling of the two reactors, irradiated fuel dispatch in
flasks to BNFL for reprocessing and the transfer of irradiated
fuel to the store. Experience to date has been very good,
bearing in mind the novel design of the plant. Breakdowns
have been relatively few since it has been in operation. Two
breakdowns of items of plant have occurred which involved working on contaminated equipment:-

1.   Faults on the chute indexing mechanism which occurred on
      that part of the mechanism within the cell. This

involved withdrawing the internal mechanism for repair and modifications.

This has occurred on 4 occasions with varying amounts of irradiated fuel in the store from 2,500 elements to 8,000 elements (early in December 1979, end of January 1980 and twice in March 1980). The contract routine maintenance equipment, ie, shielding and withdrawal equipment, designed for routine maintenance of the indexing mechanism was used and during each of the above incidents, radiation doses to operators were minimal 0.01 mSv/hr (1 mr/hr). Also, bearing in mind this mechanism is in the most active part of the route, the contamination levels on the mechanism were suffic-iently low to enable C2 working to be carried out. 37-370 Bq/cm^2 (10^{-3} to 10^{-2} uc/cm^2).

2. The second item of plant breakdown occurred on the transfer machine fuel element hoist (April 1980). The hoist drive shaft seized with a considerable amount of contaminated rope paid out. Again, radiation levels during the recovery exercise were minimal, 1.0 mSv (100 mr/hr) at the through hole of the machine but only less than 0.001 mSv/hr (0.1 mr/hr) to the operator. Contam-ination levels were low enough to enable C2 working to be carried out during unroping of the hoist and remedial work of the hoist drive.

7. General Operating Conditions

All operational areas of the plant at pile cap are classified as R2, C2, ie, the same as the reactor pile cap. Operational areas such as the switchgear room, pipe room, fan and filter rooms, where monitoring of the plant conditions are carried out on a routine basis, are uncontrolled zones. The east sprocket room is classified as an R2 C2 area only because access is via the reactor pile cap which is an R2 C2 area. The west sprocket room, which has the same access as the plant room is also an uncontrolled zone. In these areas, the general radiation level from the cell is 0.0002 mSv/hr (0.02 mr/hr).

Activity of the circulating air circuit is very low, typical values of gaseous and particulate activity being less than 0.37 Bq/ml (10^{-5} uc/ml) and less than 0.37 Bq/m^3 (10 pc/m^3) respectively. Also, a typical value of particulate activity being discharged to atmosphere from the continuous exhaust system is less than 0.037 mBq3m (1 pc/m^3) (of the order of background activity level).

Typical environmental conditions within the store (full) are:-

Air inlet temperature to heat exchangers	36°C
Air outlet temperature from heat exchangers	32°C
Relative humidity	28%
Cell depression	25 m bar below atmospheric
Cooling water flow to heat exchangers	4.5 l/s
Total power requirement for plant	9.0 KW

(fans 73 KW, exhauster 4.5 KW, dehumidifier 4.5 KW and PLC's 8 KW).

Experience to date with Wyfla Cell 4

1) Activity levels within the store circulatory system and that discharged to atmosphere are at background levels.

2) Radiation levels to operators during normal operation
    are extremely low 0.0002 mSv/hr (0.2 mr/hr).

3) There is no generation of radioactive slurries, sludge,
    ion exchange resins, etc, which have to be stored and
    eventually disposed of. The small capacity combustible
    filters have been in operation since the cell was
    commissioned and have not yet picked up significant
    contamination.

4) Radiation levels to personnel during maintenance of the
    most active part of the plant were 0.01 mSv/hr (1 mr/hr)
    with correspondingly low contamination levels. Mainten-
    ance requirements after the initial 'setting to work'
    period are minimal.

5) The new store operates at a depression with respect to
    atmospheric pressure and therefore any leakage is
    inwards. The level of depression gives accurate and
    continuous indication of integrity of containment.

6) The storage period in the store is almost indefinite.

8. What is the Significance of Being Able to Store Irradiated
    Fuel for Medium/Long Periods? (Figure 13)

Whilst the design, manufacture and construction of the large
air cooled dry stores at Wylfa was being carried out, it was
realised in the UK that these principles of dry storage
already proven for many years at Wylfa for magnox fuel, could
also be applied to irradiated oxide fuel. It was therefore
decided at that time to invest very significant time and
effort in a development programme to demonstrate feasibility
of this proposal.

Since that time, much work has been done and feasibility, cost
and schedule are now well established.

The option of long term storage of irradiated fuel allows a
more flexible approach to the question of ultimate disposal
of the irradiated fuel. In particular, the timing and type
of ultimate waste disposal can be optimised in the light of
the intrinsic value of the products from the reprocessing
cycle and the cost of reprocessing itself. The cost of
reprocessing may also reduce due to the capability of
optimising the rate of throughput of irradiated fuel through
the reprocessing cycle which a long term storage capacity
allows.

8.1 What Type of Medium/Long Term Store? (Figure 14)

As the need to store irradiated fuel for medium/long
periods is becoming established, it is necessary to
determine what type of storage best suits the need.

The authors have attempted to list the basic parameters
that they consider should be used to measure the adequacy of the many storage options currently being considered around the world.

1. Safety-Tolerance of the Design Concept to Meet the
    Latest Safety Standards

    Safety standards to which nuclear installations are
    designed vary throughout the world. Although the
    safety standards that were set for established

- 198 -

designs are extremely high, the various accident conditions that have to be considered continue to increase in number and severity. It is likely that this will be the trend in the future and it is therefore necessary to consider the mechanism of failure of new designs when the severity of the postulated hazard is increased beyond that dictated by present standards. The desirable goal must be that any failure will occur by a process of gradual degradation rather than in a catastrophic mode.

2. Dose Rate

Ability of the design of the fuel store to meet the necessary functional requirement with the minimum of additions to the station total man-rem commitment is the objective. With the pond method of fuel storage the water shielding can be by-passed because possible corrosion products on the surface of the pond may 'shine' directly to operators. This is especially true for magnox clad fuel and the problem exists to a lesser degree for any other type of fuel, especially during handling operations.

Utilisation of the dry storage principle enables storage facilities to be designed in which shielding cannot be short-circuited in this way. Although it can be demonstrated that well maintained ponds will keep dose rates down to 0.02 to 0.03 mSv/hr (2 to 3 mr/hr) it can also be demonstrated that of the total station man-rem commitment, the 'back end' of the station is responsible for a high proportion of this operator dose. It must therefore be prudent to consider carefully designs that can reduce this increment to negligible proportions.

3. Reliability

There is very strong preference for passive cooling systems that rely upon the natural heat transfer methods of convection, radiation and conduction for their safe operation. Use of an air environment for the irradiated fuel, maintained at a slight negative pressure, allows the use of containment that although nominally leak-tight, does not need absolute standards of leak-tightness to maintain established safety standards criteria. Such a design must be less vulnerable to regulatory trends.

4. Waste Generation

Unnecessary generation of radioactive contaminated material is both expensive and undesirable. The various systems must be carefully evaluated and preference given to the selection of system and processes that generate the very minimum of radioactive waste and contaminated materials. This evaluation must include examination of water treatment facilities, storage and eventual disposal of radioactive resins and sludges, decontamination and the possible need to can fuel assemblies.

5. Density Factor

In order to obtain the best utilisation of land, a high density factor is desirable in a fuel store. When comparing the density factors for different methods of storage it is obviously necessary to include all the support systems for the system being considered together

with any waste storage requirements that may be
necessary. Very often density factor comparisons are
made purely on a capacity per storage area basis rather
than a capacity "per complex" basis.

6.   Experience

It cannot be denied that to date there is far greater
experience in the pond system of irradiated fuel storage
than any other form of storage. However, it is
necessary to take into account the fact that the early
design of pond installations took place before the
design parameters for irradiated fuel storage facilities
became so exacting. It is not suggested for one moment
that pond-type installations cannot be built to the
necessary safety standards but rather that when considered against the latest design specifications, dry
storage is likely to give a better overall optimisation
of requirements.

7.   Staffing

It has been demonstrated in Britain that the minimum
number of operators for a dry storage system is typically
one third of the number required for a similar capacity
pond installation.

8.   Fuel Examination

World-wide there is more experience of storing irradiated fuel for short and medium periods in water than in
any other type of storage system. For the long term
storage of fuel however, there is little experience with
any system of storage. It is necessary therefore to
examine the design principles most carefully and to be
satisfied that, theoretically, the design should be
adequate.

However, as the ageing process is most difficult to
simulate, it is considered necessary that whichever type
of storage system is employed, means should be provided
in order that the fuel may be examined with a routine
and frequency to be determined by experience, in order
to confirm that no long term degradation trends are
present. Dry storage enables easy examination to take
place without significant increases in total dose rate
to the operator.

9.   Radioactive Release

The release of contaminated gas/gaseous fission products
to the atmosphere is obviously undesirable.

10.   'Module-Ability'

The ease with which further storage modules can be added
is a distinct advantage. Capacity of storage systems
can be tailored to closely follow the requirements of an
operational site.

11.   Schedule

The programme for design, manufacture and construction of
the first large air-cooled dry store at Wylfa has demonstrated an overall time-scale of three years for this
type of installation. The second store to a similar

design has been completed in two years. This performance is considered to be ample evidence that dry storage installations for this type and other types of fuel
and radioactive waste can be completed within three
years. The modular form of construction also enables
a phased commissioning of equipment.

12.   Costs

When carrying out cost comparisons for different
systems, it is essential that the systems being considered are designed against the same functional specifications. It is also necessary to include all costs
associated with that system in the comparison. For
example, if a certain storage system requires the fuel
to be canned then the cost of the cans, the cost of the
canning facility (or a proportion of the facility if it
is common to several stores) and the cost of the eventual disposal of the cans, must be included. Similarly,
when a water treatment plant is involved, in addition
to the cost of the treatment plant, it is also necessary
to include the cost of storage and disposal of the
spent filter resins, etc. The fact that it is difficult
to obtain these more diffuse costs is no excuse for not
including them in any comparison.

FIGURE 1 SIZEWELL FUELLING MACHINE

FIGURE 2 TRAWSFYNYDD FUELLING MACHINE

FIGURE 3 HARTLEPOOL/HEYSHAM FUELLING MACHINE

FIGURE 4   WYLFA TRANSFER MACHINE

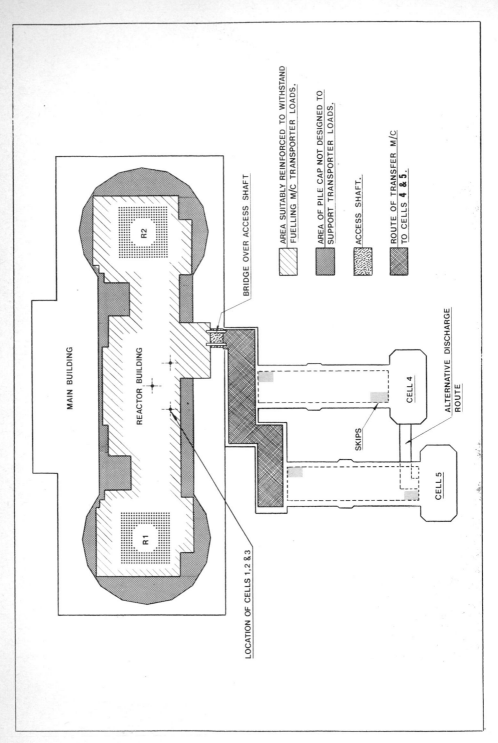

MAIN BUILDING

REACTOR BUILDING

R2

R1

BRIDGE OVER ACCESS SHAFT

LOCATION OF CELLS 1, 2 & 3

AREA SUITABLY REINFORCED TO WITHSTAND FUELLING M/C TRANSPORTER LOADS.

AREA OF PILE CAP NOT DESIGNED TO SUPPORT TRANSPORTER LOADS.

ACCESS SHAFT.

ROUTE OF TRANSFER M/C TO CELLS 4 & 5.

CELL 4

CELL 5

SKIPS

ALTERNATIVE DISCHARGE ROUTE

FIGURE 5 LOCATION OF CELLS ON CHARGE FACE

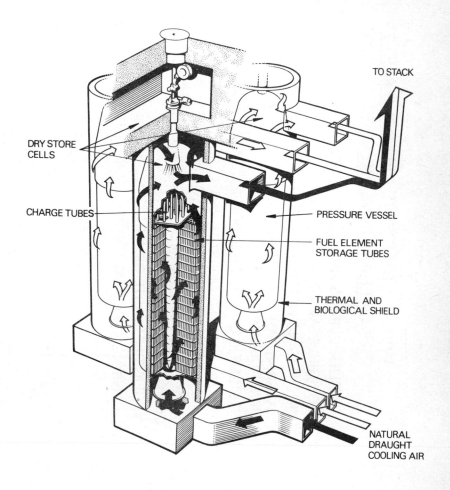

TO STACK

DRY STORE
CELLS

CHARGE TUBES

PRESSURE VESSEL

FUEL ELEMENT
STORAGE TUBES

THERMAL AND
BIOLOGICAL SHIELD

NATURAL
DRAUGHT
COOLING AIR

FIGURE 6 SECTION THROUGH CELL

WYLFA DRY STORE CELLS 1, 2 AND 3
OPERATIONAL EXPERIENCE

1 COMPLETELY PASSIVE COOLING SYSTEM IS INHERENTLY RELIABLE.

2 NEGLIGIBLE MANNING REQUIREMENT WHEN CELLS ARE LOADED.

3 WIND HAS NO DETECTABLE EFFECT ON STORE PERFORMANCE.

4 MATERIALS OF CONSTRUCTION HAVE BEEN DEMONSTRATED TO HAVE ADEQUATE LIFE.

5 NO SIGNIFICANT CONTRIBUTION TO STATION MAN-Sv COMMITMENT.

6 ACTUAL FUEL TEMPERATURES LESS THAN THOSE PREDICTED THEORETICALLY DURING DESIGN PROCESS.

7 IRRADIATED FUEL STORED FOR SIGNIFICANT PERIODS IN PRISTINE CONDITION.

8 NEGLIGIBLE WASTE GENERATION.

9 LICENSING APPROVAL FOR RELOADING INTO REACTOR IF DESIRED.

FIGURE 7 OPERATIONAL EXPERIENCE

WYLFA DRY STORE CELL 4
BASIC DESIGN CRITERIA

1 AIR TO BE THE CELL ENVIRONMENT FOR THE STORED FUEL.

2 AIR TO BE MAINTAINED AT A SLIGHT DEPRESSION WITH RESPECT TO ATMOSPHERE TO PROVIDE THE SECOND BARRIER OF THE TWO BARRIER CONTAINMENT CONCEPT.

3 MAXIMUM FUEL ELEMENT TEMPERATURE DURING NORMAL OPERATION LESS THAN 150°C.

4 MAXIMUM FUEL ELEMENT TEMPERATURE DURING WORST CREDIBLE FAULT NOT TO EXCEED 200°C.

5 NO CONDENSATION UNDER LOWEST CREDIBLE TEMPERATURES.

6 DURING NORMAL OPERATIONS MOISTURE LEVELS TO BE LESS THAN 50% RELATIVE HUMIDITY.

FIGURE 8 DESIGN CRITERIA

WYLFA POWER STATION — SCHEMATIC ARRANGEMENT OF FUEL ROUTES.

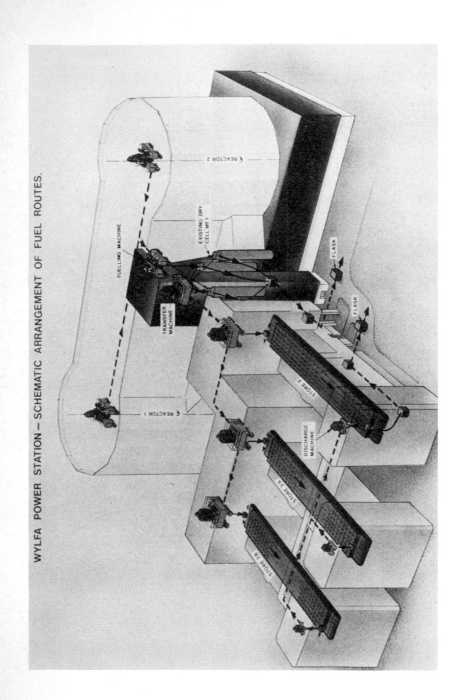

FIGURE 9   PERSPECTIVE OF CELLS 4 AND 5 SHOWING LOCATIONS WITH
RESPECT TO REACTORS.

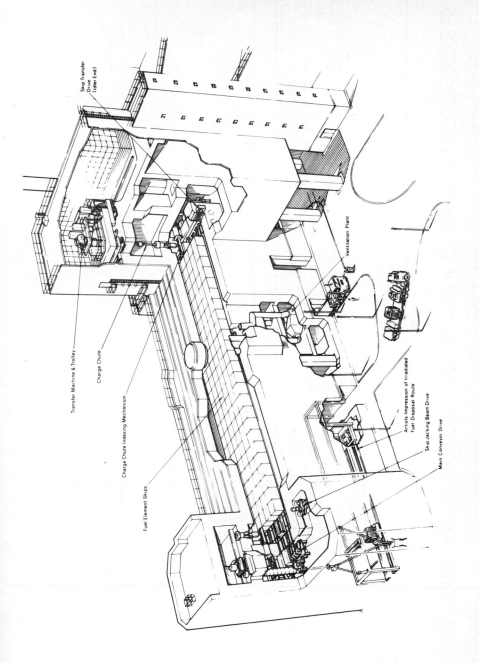

Skip Transfer Drive (Idler End)

Transfer Machine & Trolley

Charge Chute

Charge Chute Indexing Mechanism

Fuel Element Skips

Ventilation Plant

Artists Impression of Irradiated Fuel Disposal Route

Skip Jacking Beam Drive

Main Conveyor Drive

FIGURE 10 PERSPECTIVE THROUGH CELL 4.

TIME SCHEDULE FOR PLANNING AND CONSTRUCTION OF WYLFA DRY STORES

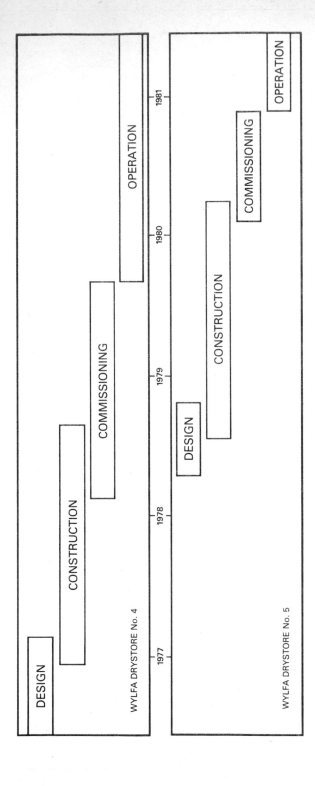

WYLFA DRY STORE CELL 4
OPERATIONAL EXPERIENCE

1. ACTIVITY LEVELS WITHIN STORE CIRCULATING SYSTEM ARE AT BACKGROUND LEVEL.

2. RADIATION LEVELS TO OPERATORS DURING NORMAL OPERATION ARE EXTREMELY LOW ~ 0.0002 mSv/h.

3. NEGLIGIBLE WASTE GENERATION.

4. MAINTENANCE REQUIREMENTS MINIMAL.

5. LEVEL OF DEPRESSION GIVES ACCURATE INDICATION OF INTEGRITY OF CONTAINMENT.

FIGURE 12 OPERATIONAL EXPERIENCE OF CELL 4.

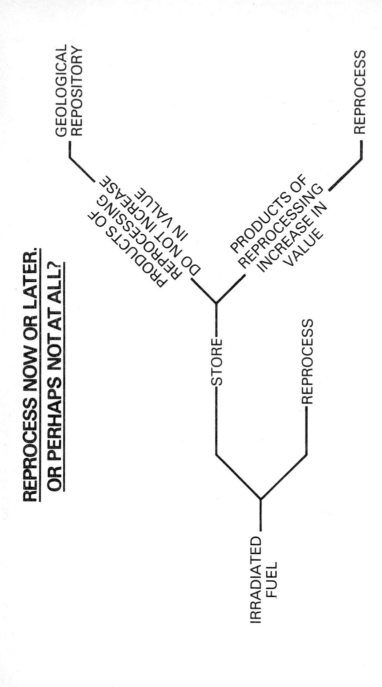

REPROCESS NOW OR LATER.
OR PERHAPS NOT AT ALL?

GEOLOGICAL
REPOSITORY

PRODUCTS OF
REPROCESSING
DO NOT INCREASE
IN VALUE

PRODUCTS OF
REPROCESSING
INCREASE IN
VALUE

REPROCESS

STORE

REPROCESS

IRRADIATED
FUEL

FIGURE 13   WHAT IS THE SIGNIFICANCE OF BEING ABLE TO STORE IRRADIATED FUEL FOR MEDIUM TO LONG PERIODS?

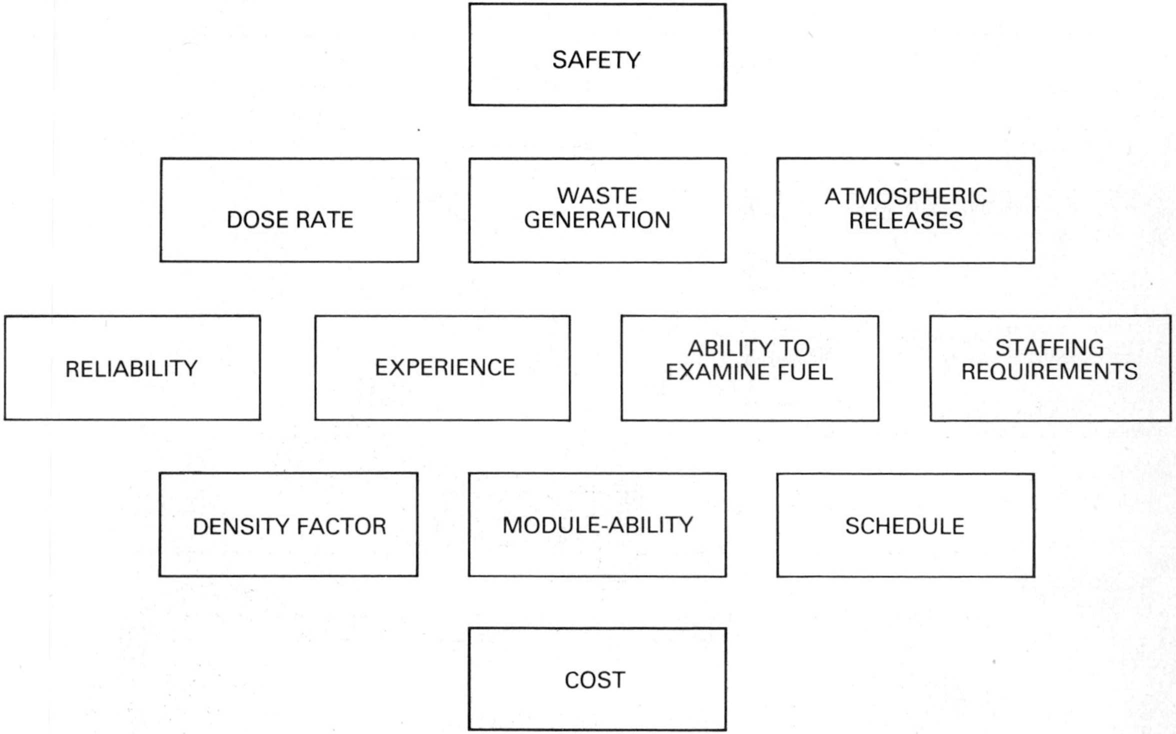

AUTHORS' VIEWS ON PARAMETERS THAT
SHOULD BE USED TO MEASURE
ADEQUACY OF THE VARIOUS
STORAGE OPTIONS

FIGURE 14

AUTHORS' VIEWS ON PARAMETERS THAT
SHOULD BE USED TO MEASURE
ADEQUACY OF THE VARIOUS STORAGE
OPTIONS.

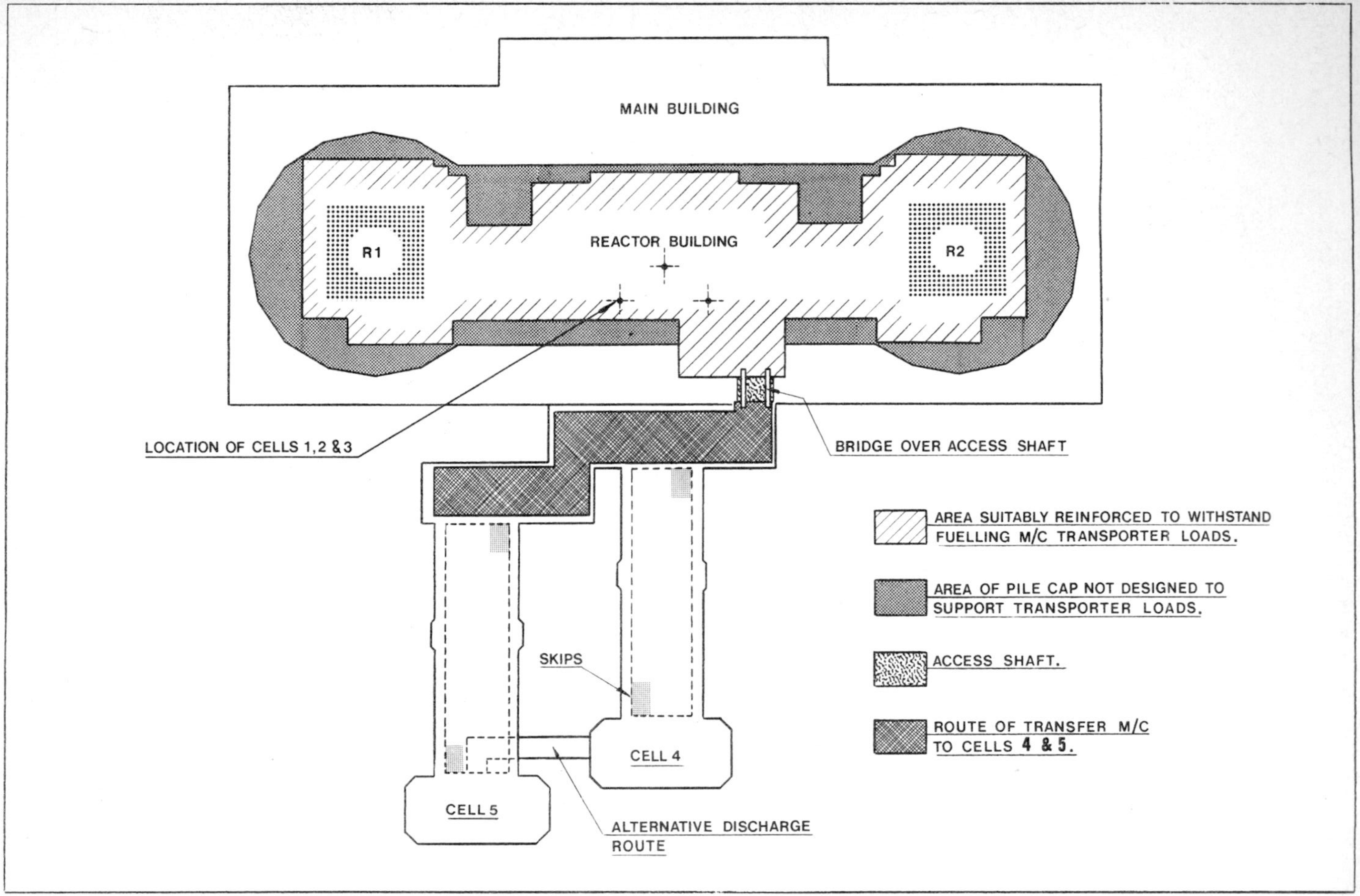

FIGURE 15 WYLFA WORKING AREA FOR FUELLING MACHINERY.

DISCUSSION

C.J. Ospina, Switzerland

How "passive" are your passive cooling-systems in your dry storage concept, bearing in mind that its normal operation requires a high pressure air atmosphere and water-cooled heat exchanger in order to evacuate the storage heat and thus guarantee that fuel canning temperatures are kept below design limits ?

E.O. Maxwell, United Kingdom

Although ours is a passive cooling system insofar as the heat transfer from the fuel element to the cooling air is concerned, the bulk heated air is cooled by forced circulation through heat exchangers. In the event that the circulating fans lose their normal electrical supply, a guaranteed supply is obtained from the back-up gas turbines automatically within a few minutes. However, we have some 8 to 9 hours in hand before fuel element temperatures reach the uranium ignition temperature limit of 275°C (assuming we have some bare uranium in the store).

C. Melches, Spain

(a)   It is obvious that the WYLFA dry storage facility (store) can be passive. What was the reason for installing intermediate cooling circuits ?

(b)   Are there plans for similar IF (irradiated fuel) stores at other nuclear stations ? If so, would they be designed on a purely passive cooling mode ?

E.O. Maxwell, United Kingdom

Due to the constraints of constructing a dry store at an operating station such as WYLFA as mentioned in the paper, we had to go for intermediate cooling circuits. If we had started, at "a green field site" then we could have engineered a totally passive system. There are no plans for similar dry stores at the Board's other magnox stations. We do not see a need for this at the moment or in the future.

H. Konvicka, Austria

What do you do with failed fuel ?

E.O. Maxwell, United Kingdom

As our reactors are fuelled on load we have a continuous burst can detection system which is monitoring each channel as the fuel is discharged from the reactor. If a failed fuel element is detected, it is treated separately i.e. canned and sent off site either to BNFL Sellafield or BNL for inspection. So, the fuel stored in the air cooled dry store is intact fuel.

STATUS OF DRY STORAGE OF
IRRADIATED FUEL IN CANADA

T. Tabe
Atomic Energy of Canada Limited
Whiteshell Nuclear Research Establishment
Pinawa, Manitoba, Canada

ABSTRACT

The irradiated fuel bundles discharged from Canada's nuclear power reactors are stored in water-pools. Air-cooled dry storage has been investigated as
an alternative to water-pool storage. Conceptual design studies of away-from-
reactor concrete canister and convection vault storage facilities for irradiated
CANDU fuel were carried out by Atomic Energy of Canada Limited (AECL) and Ontario
Hydro, respectively.

In 1974, AECL began the development and demonstration of the concrete
canister concept. Several concrete canisters were constructed for demonstration
purposes as well as for storage of irradiated fuel. In 1978, a program was begun
to assess the behaviour of fuel in a hot, dry environment.

Other uses of concrete canisters, such as transport flasks for irradiated
fuel and shielding for waste management experiments are reviewed.

1.   INTRODUCTION

        Canadian nuclear power reactors are heavy-water moderated and fuelled with
natural uranium. The irradiated fuel bundles are stored in water pools at the
reactor sites. In the early 1970's, the projection of a growing inventory of irradiated fuel justified studies for the optimization of interim storge.

        A Committee Assessing Fuel Storage (CAFS) was set up with representation
from Canadian utilities and Atomic Energy of Canada Limited (AECL). CAFS[1]
assessed two versions of water pools, those at the reactor site and those at an
independent site, and three dry storage schemes, namely concrete canisters, convection vaults and conduction vaults. One of the committee's recommendations was
to develop an air-cooled interim storage method as an alternative to water-pool
storage. Although storage of irradiated fuel in water pools has been trouble free
in Canada, the pools require continuous operation and maintenance, and they produce secondary wastes, generally from filters and ion-exchange columns in the
clean-up system. Air-cooled storage facilities can minimize these difficulties.
A number of conceptual design studies of away-from-reactor concrete canister and
convection vault storage facilities for irradiated CANDU fuel have, therefore,
been carried out, the most recent being in 1981.

        Several areas meriting further investigation were identified during the
conceptual design studies. These included methods for fuel handling and packaging, the effects of thermal stress on the concrete and the behaviour of defected
fuel in the centre of the canister at elevated central temperatures (150-200°C).
Shielding effectiveness, safeguards, safety and economic viability were also considered to be important. In 1974, AECL began a program to develop and demonstrate
the concrete canister concept[2]. Several concrete canisters were constructed
for demonstration purposes, as well as for storage of irradiated fuel from the
experimental reactor at the Whiteshell Nuclear Research Establishment (WNRE). In
1978, a program was begun to assess the behaviour of fuel in a hot, dry environment. Since the demonstration canisters were designed with security and safeguards in mind, thus making regular access difficult, and since the fuel temperature fluctuates with the outside temperature, specially designed canisters had to
be constructed to carry out this assessment in controlled conditions.

        The use of concrete canisters for purposes other than passive fuel storage
is also being investigated. Drop testing of models of the concrete canisters was
carried out to investigate their possible use as transport flasks for irradiated
fuel. In the Nuclear Fuel Waste Management Program, experiments using high-
radiation sources are planned to assess the behaviour of the various barriers in
the final disposal system. Concrete canisters will provide the shielded facilities for these experiments.

2.   CONCEPTUAL DESIGN STUDIES

        In 1975/76, conceptual design studies for storage of irradiated fuel in
concrete canisters[3] and convection vaults[4] were carried out by AECL
and Ontario Hydro in response to the CAFS recommendation. The studies were based
on a central storage facility to store all the irradiated fuel discharged in
Canada to the year 2000.

        In 1981, a study[5] was carried out based on 600 MW CANDU reactors
with a total generating capacity of 18 600 MW. The reactors were assumed to be
placed into service during the period 1989 January to 2000 January. The irradiated fuel bundles would be stored for a minimum of 10 years at the reactor sites
and would then be shipped to a central dry storage facility. The normal reactor
life was taken as 30 years, and the quantity of irradiated fuel discharged contained approximately 80 Gg U.

        Dry storage scenarios employing convection vaults (Figure 1) and concrete
canisters (Figure 2) were developed. A single central storage facility, located
an average distance of 1000 km from the reactor site, would accommodate the total
quantity of irradiated fuel discharged. The in-service date of the storage facility would be the year 2000 with retrievable storage planned to the year 2060.

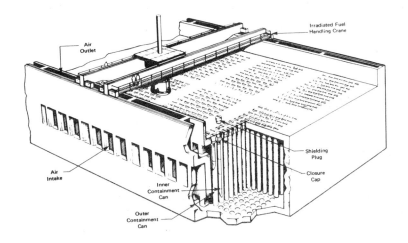

FIGURE 1:   CONVECTION VAULT FACILITY

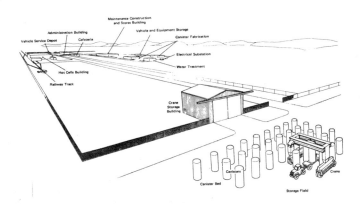

FIGURE 2:   PERSPECTIVE - CANISTER SITE PLAN

There are many similarities between the two concepts. Both would utilize natural air circulation and shielding would be provided by standard concrete. Little effort would be required to maintain the facilities. The convection vault containment can and the canister can would hold the same number of fuel bundles and have the same fuel bundle orientation, and both concepts would provide double barriers against release of radioactivity to the environment.

If convection vaults were used, the flask would arrive by rail and be placed in a water pool for unloading of the full shipping modules. The flask would then be loaded with empty modules and returned to the reactor site. The fuel bundles would be transferred from the shipping modules to storage baskets, and then to a hot cell where they would be dried with air. Six baskets would be placed in one inner containment can, which would be seal welded and inspected.

A fuel handling crane would transport the inner containment can to its storage location. Once there, the crane would remove the storage-tube shielding plug, lower the can into the tube and replace the plug. When a storage tube is full (four cans), a closure cap would be placed on the tube and seal welded. The use of seals on both the inner can and the storage tube would provide a double barrier against the release of radioactivity. Heat removal would be by natural air convection.

Each convection vault would contain 704 storage tubes with a capacity of 864 bundles, for a total vault capacity of 608×10^3 bundles or 11.2 Gg U. Eight vaults would be required to store all the irradiated fuel.

The concrete canister storage facility would be a self-contained facility with its own concrete canister fabrication plant, plant services, shipping-flask-unloading and canister-loading facilities, and a storage field for the filled canisters. The shipping flask would arrive on site by rail and be received in one of the receiving bays under the flask-unloading cell. The full shipping modules would be unloaded, and the flask loaded with empty modules and returned to the reactor site.

The fuel bundles would be transferred to the canister basket. When the basket is full a cover would be installed, seal welded and inspected. Three sealed baskets would be placed in the canister can which would have its cover installed, seal welded and inspected.

An empty concrete canister would be positioned under the canister-loading cell and loaded with a full sealed can. The canister plug would be installed and seal welded, and the depressions grouted. The canister would then be transported to the canister storage field.

The canister storage field would consist of seven beds, each capable of storing 3240 canisters arranged in 108 rows of 30 canisters each. Each concrete canister would hold 216 fuel bundles. Twenty thousand concrete canisters would be required to store all the irradiated fuel.

The total estimated storage costs, expressed in 1980 July 1 Canadian dollars, were calculated by summing transportation, capital and operating costs. Since transportation costs are the least well defined, the sensitivity of total cost to variations in transportation costs was evaluated by considering three different freight rates. The total unit costs incurred in implementing the convection vault storage facility and the concrete canister storage facility are shown in Table I. In all cases, the transportation cost accounted for more than 50% of the total cost.

3.   CONCRETE CANISTER DEMONSTRATION PROGRAM

In 1974, AECL initiated a program to develop and demonstrate the concrete canister concept. Four canisters were designed and constructed at WNRE, two for electrical heating and two for loading with irradiated fuel. One of each is cylindrical, with the design based on the reference concept; the other is essentially square in cross section with the outer corners trimmed to enhance

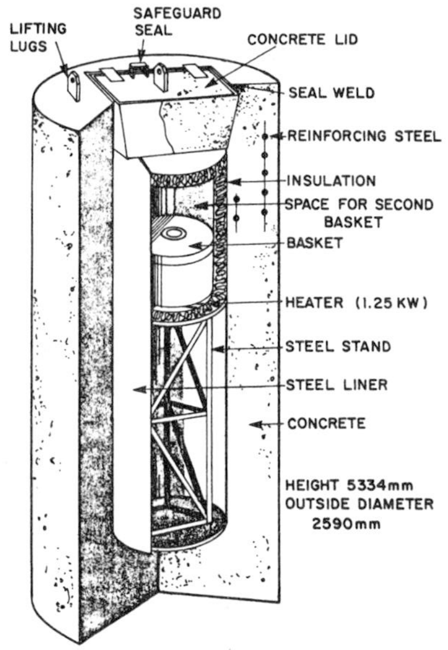

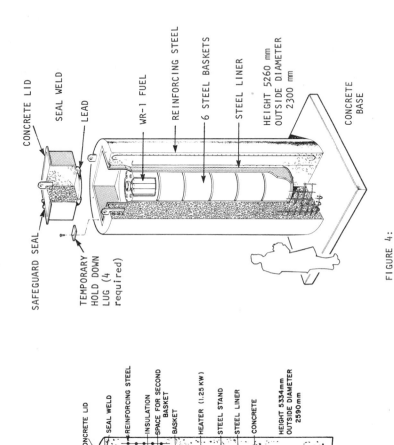

CONCRETE LID

SEAL WELD

LEAD

SAFEGUARD SEAL

TEMPORARY HOLD DOWN LUG (4 required)

WR-1 FUEL

REINFORCING STEEL

6 STEEL BASKETS

STEEL LINER

HEIGHT 5260 mm
OUTSIDE DIAMETER 2300 mm

CONCRETE BASE

FIGURE 4:
OPTIMIZED CYLINDRICAL CONCRETE CANISTER

SAFEGUARD SEAL

CONCRETE LID

SEAL WELD

REINFORCING STEEL

INSULATION

SPACE FOR SECOND BASKET

BASKET

HEATER (1.25 KW)

STEEL STAND

STEEL LINER

CONCRETE

HEIGHT 5334mm
OUTSIDE DIAMETER 2590mm

LIFTING LUGS

FIGURE 3:
CONTROLLED ENVIRONMENT EXPERIMENTS CANISTER

structural integrity. Testing of the electrically heated cylindrical canister began early in 1975. Fuel from the WNRE research reactor (WR-1) was loaded into the
second cylindrical canister during the fall of 1975. The electrically heated
square canister became operational early in 1976, and a square canister was loaded
with fuel from Douglas Point Generating Station in Ontario in mid-1976.

The electrically heated cylindrical canister was tested at power levels
ranging from 0-500 percent of design (0-10 kW). The first hairline cracks were
observed at 1.5 - 2.0 kW, as anticipated, but the cracks did not affect the
structural or shielding integrity of the walls, even at higher power. The electrically heated square canister was tested at power levels ranging from 0-11 kW.
Its performance was also satisfactory. The heated cylindrical canister was subjected to over 50 freeze/thaw cycles without showing visible deterioration of the
surface. Both calculations and tests have given conclusive evidence that the
present canister design has a wide margin of safety in terms of thermal loading
and stress. Other aspects of canister acceptability, such as shielding effectiveness, man-rem requirements, safety analyses, safeguards and transportability, were
also satisfactory.

Because of their excellent performance, WNRE has opted to use canisters to
provide additional storage capacity for irradiated WR-1 fuel. Seven cylindrical
canisters have been built at the WNRE waste management area. Three of the canisters are used for experimental purposes (Figure 3). The remaining four are used
for storage of WR-1 fuel (Figure 4), although one basket in one of these storage
canisters is used for a fuel storage experiment under dry conditions at seasonally
varying temperatures. Three of the storage canisters were constructed during 1977
and 1978, and were loaded with fuel in 1977 July, 1978 October and 1980 November,
respectively. Two canisters to house experiments were constructed in 1980 and
loaded in 1980 October and 1981 May. Two canisters, one for storage and one for
experiments, were constructed in 1981. The experimental canister was partially
loaded in 1981 November, with the final loading to be completed in 1982 June. Two
baskets were loaded in the storage canister in 1981 November. The remaining four
baskets are scheduled for loading starting in 1982 May. One storage canister is
scheduled for construction in the summer of 1982 and loading is scheduled to start
in the fall of 1982.

Since Canada is a signatory to the Non-Proliferation Treaty, all nuclear
material is subject to International Atomic Energy Agency (IAEA) safeguards.
Therefore, a safeguards procedure was initiated for the WNRE canister program.
This involves surveillance of the fuel bundles by an IAEA inspector from the initial transport stage to the final loading. The procedure uses bundle identification, temperature measurements, and gamma radiation measurements to give satisfactory assurance of the basket contents. Once a canister is loaded, two IAEA seals
are applied to the lid.

Several experimental programs associated with canister safeguards have
been undertaken. One used thermoluminescent devices on the canister surface to
yield information on the contents. The devices were sensitive enough to indicate
the presence or absence of a large quantity of bundles, but could not detect a
single missing bundle. It was found that equivalent information could be obtained
more easily and quickly using an ion-chamber-type survey meter.

Another experiment involved the testing of fibre-optic seals. The fibre-
optic cables were passed through a U-tube, embedded in the canister body and lid,
and then located along the canister side for viewing from ground level. In late
1977, two fibre-optic seals were applied to one canister, in addition to the
standard IAEA seals. In late 1979, another set was installed on a second canister. The second set was connected to the Remote Continual Verification (RECOVER)
system, which allows inspection of the seals from IAEA headquarters in Vienna,
using a telephone. The verification of the fibre-optic seals presented no major
problems; however, the cold weather caused the outer plastic covering on the cable
to shrink, causing the seals to pull apart. The RECOVER system lacked continuity
of information due to problems with transmission to Vienna. It was agreed that
the technology was not yet well enough developed, and the program was abandoned.
The canisters remain sealed with two stainless steel wire seals, and they are
inspected approximately three times a year.

Routine inspections of the loaded storage canisters and the canister storage site are carried out on a quarterly basis. Visual inspection of the canisters
has shown no deterioration of the canister surfaces. Gamma-field surveys, on contact and 2 m from the canister, are taken at eight points around each canister and
results indicate satisfactory performance. The internal space between the canister liner and the sealed fuel basket is monitored continuously for fission products and moisture, using a closed air-circulating system. No abnormal activity
levels or signs of water in-leakage have been observed. Visual inspection, peri-
meter dosimetry and analysis of groundwater and run-off from the canister storage
site have shown no abnormal changes.

4. OTHER USES OF CONCRETE CANISTERS

4.1 Controlled Environment Experiment

A program, jointly undertaken with Ontario Hydro, was set up at WNRE to
generate experience regarding the behaviour of irradiated fuel stored under conditions pertinent to dry storage. An experiment referred to as the Controlled Environment Experiment (CEX) was devised. It utilizes concrete canisters and carefully characterized fuel bundles. Electrical heaters, located inside the canister, are used to control the internal temperature.

The CEX consists of two phases. Phase 1 involves the storage of deliberately defected and undefected bundles in essentially dry air at 150°C. Eight bundles, four defected and four undefected, were loaded into a canister in 1980
October. Prior to storage, two elements from each bundle were removed for pre-
storage examination to define the initial conditions. A third element, designated
the "control" element, was removed from each bundle, γ-scanned for isotope distribution, visually examined and returned to the bundle to be used as a calibration
standard during subsequent examinations. This experiment is unattended except for
routine monitoring of the air temperature in the canister. The first inspection
of the fuel elements is expected to be carried out in late 1983.

Phase II of the program involves the storage of eight fuel bundles at
150°C in moist air. Half of the bundles will have their outer elements punctured
to expose the contained fuel to the humid atmosphere. Because of the unavailabil-
ity of fuel, only six fuel bundles were loaded into the canister in 1981 November.
The remaining two bundles have been received and are scheduled to be loaded in
1982 June. The first scheduled inspection of the fuel bundles will be after approximately three years of storage.

4.2 Canisters as Transport Flasks

If irradiated fuel were to be stored routinely at a special site, a large
number of flasks would be required to transport it from the reactors to this central facility. Shipment in traditional flasks would require fuel handling at both
the reactor sites and the central facility. To reduce fuel handling, the use of
concrete canisters for both shipping and storage was proposed.

The use of canisters as shipping flasks would require additional facilities at each reactor site. However, the elimination of traditional flasks and
reduced handling might result in a large net cost saving. Furthermore, the reduction in handling would reduce personnel exposure. It was agreed, therefore, that
the potential benefits merited further investigation.

A series of drop tests was carried out to help assess the ability of the
canisters to survive the tests that would be required to licence them. Two 1/8
scale models, filled with steel annuli to simulate the mass of the fuel, were
tested at AECL's Chalk River Nuclear Laboratories (CRNL). Each canister was subjected to two 9-m drops (corner and flat) and a 1-m puncture drop. The standard
IAEA impact target configuration was used for the drop test. For the puncture
tests, a 19 mm diameter steel pin (1/8" scale) protruding 73 mm was used.

Canister I was first subjected to the corner drop test, then the puncture
test on the damaged corner, and finally the flat drop test. The corner drop test

resulted in the loss of concrete in the contact corner, deformation of exposed reinforcing bar, and cracks along the axis of the canister. In the puncture test,
the pin impacted an exposed reinforcing bar which restricted its penetration and
resulted in a puncture hole only. The final flat drop resulted in a great deal of
spalling of concrete from around the outer reinforcing bars.

Canister II was subjected to a flat drop test, a flat puncture test and a
corner drop test. The flat drop caused spalling of a large area of concrete from
the outer reinforcing bars and the formation of several large cracks. These
cracks were oriented at an angle to the impulse load typical of a shear failure in
brittle material. In the puncture test, the pin struck an area with little rein-
forcement, resulting in full penetration and loss of concrete in the puncture
area. The final corner drop resulted in further loss of concrete and deformation
of the exposed reinforcing bars.

After the series of tests, however, both containers were still basically
intact. No detectable breaching of the inner steel liner had occurred. Also
there was no evidence of any liner deformation which might restrict removal of the
simulated baskets. However, the combination of tests resulted in a loss of approximately 20% (by weight) of the shielding material. In combination with the
cracks, the resulting shielding integrity would fail to meet IAEA standards.

The lack of deformation or breaching of the liner was very encouraging.
However, it was concluded that the existing canister design does not meet IAEA
requirements. Additional reinforcing bars on the bottom and changing of the
standard concrete mix to ilmenite concrete or steel fibre reinforced concrete were
recommended, to limit damage from puncture tests and to reduce spalling and
cracking. Additional testing is required to attempt to resolve the problems.

4.3   Shielding For Waste Management Experiments

The Canadian concept for the management of high level nuclear fuel waste
proposes that it be disposed of in an underground vault. The nuclear fuel waste
in the final disposal system will be protected by several barriers, which will
help prevent the transfer of radionuclides to man. Since the performance of the
total system cannot easily be demonstrated in reasonable times, the behaviour of
individual barriers must be verified. Many of the processes being investigated
are chemical in nature and will depend significantly on water chemistry. Because
radiation can affect water chemistry, a significant number of experiments must be
performed in high-radiation environments.

The experiments will include corrosion under postulated worst-case
disposal-vault conditions, studies of the performance of small-scale simulations
of the containment systems, and basic research studies on processes thought to be
important in the system.

Experimental modules will be prepared in hot cells at WNRE and then transferred to a canister-loading bay located in the new Irradiated Fuel Test Facility
presently being built at WNRE. The modules will be placed in concrete canisters,
which will provide shielding and module heating. A typical module will be 0.5 m
in diameter and 1 m high, and will be housed in a concrete canister approximately
2 m in diameter and 2 m high.

There will be two classes of experiments. The majority will be "passive"
in so far as they will not be instrumented except for temperature monitoring. All
the information derived from these experiments will be obtained after they are returned to the hot cell and stripped down for examination. The "active" experiments will require other instrumentation and will often involve perturbation of
conditions inside the module by external means. The length of an experiment could
range from a few days (for active experiments) to several years (for passive
experiments). The facility is scheduled for service in the later part of 1982.

5.   CONCLUSIONS

Canisters have more than fulfilled our expectations for irradiated fuel storage. Canisters can be designed and constructed to meet the individual requirements of the user, and they are economical and versatile.

Although the integrity of the inner liner following drop tests is encouraging, work is required to solve the problem of cracking and spalling of the canister surface before canisters can be considered further for irradiated fuel transportation.

The use of canisters as shielding for waste management experiments clearly indicates their promise for a variety of uses.

REFERENCES

1.   Morgan, W.W. (editor): "Report by the Committee Assessing Fuel Storage", Atomic Energy of Canada Limited Report, AECL-5959 (1977).

2.   Ohta, M.M.: "The Concrete Canister Program", Atomic Energy of Canada Limited Report, AECL-5965 (1978).

3.   Lidfors, E.D., Tabe, T. and Johnson H.M.: "Conceptual Design Study of a Concrete Canister Spent Fuel Storage Facility", Atomic Energy of Canada Limited Report, AECL-6301 (1979).

4.   Barnes, R.W.: "The Management of Irradiated Fuel in Ontario" Ontario Hydro Report, GP-76014 (1976).

5.   Tabe, T. Rao, P.K.M. and Lightstone, L.: "Management of Irradiated Fuel: A Description of the Dry Storage Concept for the Proposed Mexican Nuclear Program", Whiteshell Nuclear Research Establishment Unpublished Report, WNRE-513 (1982).

TABLE I

COSTS FOR FUEL MANAGEMENT

$ Canadian July 1980

Description	Freight Rate $55/Mg+$0.095/(Mg.km)		Freight Rate $0.35/(Mg.km)		Freight Rate $0.50/(Mg.km)	
	Canister	Convection Vault	Canister	Convection Vault	Canister	Convection Vault
Capital	0.49	0.89	0.49	0.89	0.49	0.89
Operating	4.19	2.10	4.19	2.10	4.19	2.10
Transportation	5.00	5.00	10.34	10.34	14.37	14.37
Total ($/kg U)	9.68	7.99	15.02	13.33	19.05	17.36

DISCUSSION

M. Peehs, Federal Republic of Germany

In civil engineering, concrete is only specified by its ultimate compression strength. However, in your application you refer to many other qualities of concrete normally not specified, for example heat conductivity, water content, type of aggregate etc. My question is : what is the specification of your concrete under the special aspect of its application as structural material of a storage canister ?

T. Tabe, Canada

During the design stage of the concrete canister, analyses have been carried out that the concrete would withstand the thermal loading and stress based on an estimated heat load of 2 kW. However, as I understand it, the canisters are built using a standard concrete mix based on the strength of the concrete.

R. Christ, Federal Republic of Germany

You have mentioned two barriers. Would you please identify these barriers ?

T. Tabe, Canada

The fuel bundles are placed in the canister basket which is sealed to form the first barrier. The sealed basksets are then placed into the canister which has a steel liner. The steel liner is then sealed to form the second barrier.

G.A. Brown, United Kingdom

I note that your vault design has side air inlets and top outlet stacks. We have been investigating air vaults and find that the effects of high prevailing winds on the building can cause a depression on the downwind side which can be of the same order as the natural convection buoyancy pressures. Cooling can therefore be seriously affected. We do not see a future in side inlets.

T. Tabe, Canada

The vault design work was carried out by Ontario Hydro. It was pointed out later that the conceptual design concentrated more on feasibility of the vault in handling the fuel and providing adequate storage capacity, and on providing comparative costs. Design of the cooling aspects was not considered in detail and this area of the wind effect was not investigated.

M.S.T. Price, United Kingdom

Is it your intention to pursue the development of concrete "canisters" or "casks" as type B transport containers ? I suggest that there is also interest in developing concrete containers for storage and eventual disposal of intermediate level waste, both operational and decommissioning wastes.

T. Tabe, Canada

The drop tests were carried out to assess if there was any chance of the canister trying to meet IAEA requirements as a type B transport canister. Tests results have indicated that further work is required.

M.S.T. Price, United Kingdom

What is the maximum possible height that a canister could be accidentally dropped during the whole canister handling route.

T. Tabe, Canada

Although the tests were carried out at the required 9 m drop, the expected height that a canister could drop would not be much more than slightly higher than the height of the transporting medium i.e. truck or rail.

AN EVALUATION OF THE USE OF CONCRETE CASKS FOR IRRADIATED FUEL MANAGEMENT

S.J. Naqvi, J. Freire-Canosa, C.R. Frost
Ontario Hydro
Toronto, Canada

and

K.M. Wasywich, M.G. Wright
Atomic Energy of Canada Ltd.
WNRE
Pinawa, Manitoba, Canada

ABSTRACT

The status of the Canadian Concrete Canister Dry Storage program is briefly reviewed.

A preliminary evaluation indicates that significant cost savings, compared to a reference concept, could result from the use of a concrete canister (or cask, as a multi-purpose canister is called) for transportation and disposal, as well as storage.

The on-going Canadian program to investigate the behaviour of defected and non-defected irradiated fuel in dry transportation and storage conditions is described. Results show that the definition of a maximum temperature for dry transportation and/or storage should be approached with caution.

1.0 INTRODUCTION

Irradiated fuel from Canadian Deuterium Uranium (CANDU) reactors is currently stored in water-filled pools at the reactor site.[1] These pools, also called irradiated fuel bays (IFB's), are of two types:

(a) Primary bays;
(b) Auxiliary or secondary bays.

The primary IFB's are an essential part of station design and provide for reactor refuelling operations as well as initial storage and cooling. The auxiliary IFB's are designed to store fuel after initial cooling in primary bays and provide for additional storage capacity as required.

Although over 20 years of storage experience with irradiated zircaloy clad fuel in water pools has been trouble free,[2] it is considered that dry storage based on passive cooling could have potential economic advantages over the present wet storage method. Several dry storage concepts have been considered in Canada including:

(a) At-surface concrete canisters or silo storage;
(b) At- or near-surface or deep underground vault storage; and
(c) At- or near-surface and deep underground drywell storage, or direct emplacement.

Of these concepts, only the concrete canister has been developed to a stage where it can be seriously considered as an alternative to water pools. The other concepts are only at the conceptual level and were further examined in a study investigating the feasibility of storing irradiated fuel for periods greater than 50 years.[3] This Ontario Hydro study evaluated various options for longer term fuel storage in the event that ultimate disposal in deep underground geological formations is deferred.

The overall status of dry storage concepts in Canada is reviewed in another paper at this conference by Whiteshell Nuclear Research Establishment (WNRE) staff[4]. In our paper, only the concrete canister concept is discussed, with an evaluation of its commercial viability in an integrated, irradiated fuel management system.

A key element in assessing the viability of this concept is the behaviour of irradiated fuel under the varying conditions in a canister when used in the various phases of an integrated irradiated fuel management system. An experimental program to develop irradiated fuel behaviour data covering these likely conditions has been initiated jointly between Ontario Hydro and Atomic Energy of Canada Limited (AECL). A status review of these experimental programs is provided in this paper.

2.0 CONCRETE CANISTER PROGRAM

The development and demonstration of concrete canisters as a viable interim irradiated fuel storage concept has largely been an AECL effort.[4] In 1974, AECL decided to develop the concrete canister concept after concluding that canisters would be cheaper to build and operate than water pools.[5]

At that time, four canisters were designed and constructed at WNRE. These canisters have since been tested with electrical heaters and with irradiated fuel from WNRE's research reactor (WR-1) and from Ontario Hydro's Douglas Point Nuclear Generating Station. Both cylindrical and square shaped canisters have been built and tested at WNRE.

The reference concrete canister is a concrete cylinder 5 m high, 2.5 m in diameter, with a 0.75 m diameter internal cavity. The storage capacity of each canister is 4.5 MgU or 216 Pickering type fuel bundles which, after 5 years of pool storage, will have thermal power of about 2 kW in the canister. The fuel is loaded into baskets, sealed in cans and placed in the central cavity. Heat transfer is mainly by convection to the inner cylinder wall, by conduction through the concrete and natural convection over the outer surface of the canister.

Assessments have been carried out to determine the effectiveness of concrete storage canisters in meeting the following requirements: shielding (minimum man-rem dose); accident safety; safeguards; transportation. Both theoretical analysis and experimental tests have given conclusive evidence that the canister, as now designed, has wide margins of safety in terms of all these factors.

The use of concrete canisters to evaluate the behaviour of defected and non-defected irradiated fuel in dry storage is described in Section 4.2 below.

3.0   CONCRETE CASKS* IN INTEGRATED IRRADIATED FUEL MANAGEMENT

An economic evaluation of an integrated irradiated fuel management system (including interim storage, transportation, immobilization and disposal) based on a concrete casks was recently carried out in Ontario Hydro.[6] The economic incentives and conceptual feasibility of using concrete casks for all back-end fuel cycle operations were determined by comparison with a reference concept for irradiated fuel management. The reference concept involves storage in waterpools at reactor site, transportation in metallic casks and finally immobilization in a lead-invested durable metal container and burial in a hardrock underground repository. Two basic types of casks were considered: (a) ordinary reinforced concrete; and (b) heavy reinforced concrete. A special polymer impregnated concrete cask design is proposed in order to achieve the long term durability required for permanent fuel waste disposal.

The multi-purpose concrete cask design (shown in Figure 1) has a capacity for 3 modules (96 bundle irradiated fuel containers) for a total of 288 bundles. Their outer shape and inner cavity is rectangular prismatic. The inner cavity is lined with carbon steel and is designed to contain the modules with a minimum of tolerance between the module and liner. The cask is also provided with a concrete closure plug lined on its side and bottom with carbon steel. It provides a tight fit with the inner cavity of the cask.

---

*   The term cask is used henceforth to refer to a specific concrete canister design with multi-purpose applications, i.e., storage, transportation, and disposal.

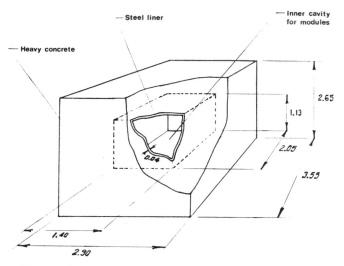

Figure 1. Heavy concrete cask. Dimensions in meters

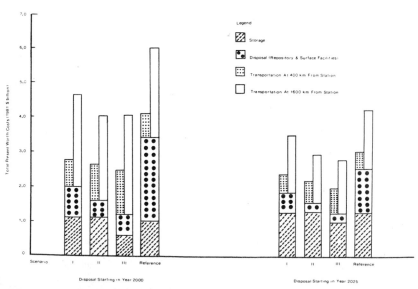

Figure 2. Comparison of Scenario Costs for the Main Components of Irradiated Fuel Management

Property	Ordinary Reinforced Concrete	Heavy Reinforced Concrete
Wall thickness (m)	0.95	0.75
Density (tons/m^3)	2.4	3.5
Interior Steel Liner Thickness (m)	0.04	0.04
Cask Weight (tons)	90	93
Cost (1981 k$)	16	23

Three options for using concrete casks in various roles in the back end fuel cycle could have potential economic advantages over the reference concept. These options are:

(i) Option I: Disposal of irradiated fuel in polymer-impregnated ordinary reinforced concrete casks.

(ii) Option II: Transportation and disposal of irradiated fuel in a single polymer impregnated heavy reinforced concrete cask.

(iii) Option III: Storage, transportation and disposal of irradiated fuel in a cask identical to (ii) above.

The economic comparisons were based on the following assumptions about the nuclear generating system:

Installed generation capacity:	62,200 MW(e)*
Station Life:	50 years
Disposal Dates:	2000 and 2025
Transportation Distances:	400 km and 1600 km
Age of fuel moved to dry storage or disposal:	5 year cooled

The costs for the three main phases of irradiated fuel management (storage, transportation and disposal) are compared for all the options in Figure 2.

As Figure 2 clearly indicates, significant overall cost reductions for irradiated fuel management result from the use of concrete casks in one or several phases of the back end fuel cycle operations. Based on a present worth cost analysis, minor savings are made when concrete casks are used for storage only. However, savings of about $2 billion and $1 billion (1981$) can be realized for disposal in 2000 and 2025 respectively, when casks are used for transportation and disposal. The option using concrete casks for storage, transportation and disposal appears to have the greatest potential for further savings in irradiated fuel management.

4.0   IRRADIATED FUEL BEHAVIOUR IN DRY STORAGE

    4.1 Background

Irradiated fuel stored in a dry environment (air) will experience more severe conditions than fuel stored in water pools. In a Canadian climate an ambient condition for the

---

* This unrealistically large generating system capacity was used for comparison purposes because if facilitated direct comparison with a previously estimated reference concept based on the 62,200 MW(e) system.

- 233 -

concrete canisters will vary from +40°C to -40°C. As the decay
heat decreases, the irradiated fuel will eventually experience
virtually the same temperature range. The higher initial
temperatures of storage (100-150°C) may increase susceptibility
to fuel sheath cracking by stress corrosion or embrittlement by
fission products such as Iodine, Cadmium and Cesium.
Similarly, higher oxidation rates of UO_2 may lead to failure
of defected fuel bundle sheaths when stored in air. While
these effects are not expected to be significant during the
envisaged periods of storage and/or disposal an experimental
program was undertaken to confirm this expectation. A joint
program between Ontario Hydro and AECL was initiated to collect
experimental data on the behaviour of defected and non-defected
irradiated fuel in dry storage. The main objectives are to
study the integrity of the sheath and bundle, and evaluate the
effect of any possible loss of integrity on long-term fuel
management. The program consists of three concrete canister
experiments and a laboratory test program to investigate fuel
sheath cracking and UO_2 oxidation.

4.2   Current Status of Dry Storage Experimental Programs

The three concrete canister experimental programs
are: a) Easily Retrievable Basket (ERB), b) Controlled
Environment Experiment Phase 1 (CEX-1), and c) Controlled
Environment Experiment Phase 2 (CEX-2). The objectives and
status of each of these three programs, and the fuel clad
cracking and UO_2 oxidation programs are described below:

a)   Easily Retrievable Basket (ERB)

The objective is to investigate the stability of
irradiated CANDU fuel bundles stored in dry air at seasonally
ranging temperatures. This concrete canister contains a total
of 159 bundles in five baskets plus the ERB. The ERB contains
two Pickering bundles, ten Whiteshell reactor (WR-1) UO_2
bundles and one WR-1 uranium carbide bundle. It was loaded in
October 1978, after pre-storage characterisation of the bundles.

The temperatures of the ERB fuel bundles have been
recorded and documented since loading. The air in the ERB is
monitored for radioactivity to detect if any of the bundles
defect during the experiment.

The ERB is scheduled for retrieval in May 1982, 3-1/2
years after initial loading. The bundles will again be
characterised to ascertain if any changes have occurred and
then reloaded. The outcome of the characterisation will
determine the timing of the next ERB retrieval.

The temperature difference between the fuel and the
outside ambient temperature was initially about 50°C.[7]
Since then, this temperature difference has dropped gradually
to about 25°C due to the decreasing bundle decay heat.

b)   Controlled Environmental Experiment (CEX) - Phase 1

The Controlled Environment Experiment (CEX) tests were
initiated to overcome the limitations of the ERB experiment.
The tests involve the storage of fuel bundles at relatively low
temperatures compared to those predicted for ultimate dry
storage. Phase 1 of this experiment includes storing eight
irradiated fuel bundles (four from Pickering NGS and four from
Bruce NGS) in hot (150°C) dry air in a modified concrete
canister. In this experiment the outer elements in half of
each type of bundle were deliberately defected prior to storage

in order to determine the effects of storage conditions on defected and non-defected fuel elements. The canister basket has two isolated compartments with defected bundles in one and non-defected bundles in the other. The basket containing the fuel bundles was loaded into the canister in October 1981.

Before the fuel bundles were loaded into the canister, three outer elements and one element from each inner ring of each bundle were removed for characterization. One of the three outer elements per bundle was a control element and was examined nondestructively before being loaded into the canister with the bundles. The remaining two outer elements on each bundle are being examined destructively. Most of the characterisation work on the Pickering and Bruce elements has been completed. The remainder of the characterization work will be completed by June, 1982.

The first interim examination for this experiment is currently scheduled for 1985.

c)   Controlled Environment Experiment (CEX) - Phase 2

In a large scale irradiated fuel storage operation, water may be transferred on the bundle surfaces or in defective elements from the water bays to the basket in which the bundles are to be stored. A high humidity in the container may have a significant effect on the degradation of the defected fuel bundles. The objective of the CEX-2 test is therefore to investigate the stability of irradiated fuel bundles under hot (150°C) moist storage conditions.

This experiment involves storing four Bruce and four Pickering irradiated fuel bundles. In this experiment, as in Phase 1, the outer elements in half of each type of bundle were deliberately defected before storing. Unlike Phase 1, these bundles were placed in individual pressure vessels charged with water, and these in turn were loaded into a canister basket for storage.

The bundles were inspected and dimensioned before storage. In addition, three elements from each bundle were removed for detailed characterization. Sections of the elements will be retained for clad crud analysis, stress corrosion cracking and metal vapour embrittlement tests, and UO_2 leaching tests.

The pre-storage examination of the fuel elements is continuing and should be complete by December 1982. A bundle characterization report is planned for February 1983.

4.3   Description of Laboratory Experiments

a)   Fuel Clad Cracking Tests

The objective of this program is to investigate the vulnerability of irrradiated fuel clad to stress corrosion cracking (SCC) by iodine or metal vapour embrittlement (MVE) by Cs/Cd at temperatures anticipated in long-term dry storage facilities. This program should provide design data on the maximum temperature at which we can dry store irradiated fuel bundles.

Work in 1982/83 will be performed at AECL's/Chalk River Nuclear Laboratory (CRNL). Any subsequent work will be at AECL's WNRE facilities. Stressed irradiated fuel clad ring specimens that have been intentionally precracked by iodine will be used in the 2000 h tests. The tests will encompass a

range of applied stress on the rings at temperatures from 100°C
to 300°C with different iodine or cesium/cadmium vapour
pressures. Any future work at WNRE in this program will be
initiated after the results from the CRNL tests in 1982/83 have
been evaluated.

Results to date in similar experiments to those
described above have shown:[2]

i)   With iodine at a vapour pressure of 36 Pa, the
critical stress intensity (K_{ISCC}) to cause SCC
was virtually constant at about 9 MPa $\sqrt{m}$ between
20°C and 100°C, but dropped sharply to 4
MPa $\sqrt{m}$ over the temperature range 100°C to
150°C. K_{ISCC} fell to about 3 MPa $\sqrt{m}$ in the
range of 150°C to 200°C. The threshold
starter crack in these tests was about 40%
through the clad.

ii) Cesium and cadmium appear to act synergistically
to cause clad embrittlement. There is a sharp
drop (from about 10 MPa $\sqrt{m}$ to 3 MPa $\sqrt{m}$) in the
critical stress intensity (K_{IMVE}) in the
temperature range of 100°C to 150°C under the
test conditions (the cesium and cadmium vapour
pressure at the temperature of the experiment).

b) UO_2 oxidation

Hydro has several joint on-going programs on UO_2
oxidation at AECL/CRNL. We are currently reviewing the results
from these programs and from the literature. If any further
work is required to evaluate aspects of UO_2 oxidation
relevant to dry storage, it will be initiated as part of the
dry storage program.

Latest oxidation results[8] indicate irradiated
UO_2 in defected elements should not cause clad splitting for
periods up to at least 500 h at 230°C and up to 700 h at
220°C.

Very little published data exists that can be used to
obtain a quantitative measure of the rate of oxidation of
defected irradiated fuel at temperatures of interest in dry
storage (between 150°-250°C). From the available data it is
difficult to determine the temperature at which UO_2 to
U_3O_8 conversion begins and how oxidation affects cladding
degradation.

Ainscough and Oldfield[9] studied oxidation of
un-irradiated defected UO_2 fuel elements at 175°C in 1964.
They observed that the weight gain per element was less than 10
mg in 1440 hours. This corresponds to an increase of O/U ratio
of less than 0.001. Their metallographic examination indicated
the occurrence of oxidation close to the defect. No evidence
of lower density phases, such as U_3O_8, was found. Boase
and Vandergraff[10] examined UO_2 to U_3O_8 oxidation in
somewhat greater detail. Their work shows that at 200°C
defected irradiated fuel elements could oxidize enough to
eventually split the cladding. Extrapolation of oxidation rate
vs temperature in their paper indicates that complete oxidation
of an element at 200°C would take 48-96 years. However, clad
splitting has been observed by other experimenters[9] at
250°C at an average O/U ratio of 2.05 which corresponds to 7.5%
oxidation to U_3O_8. Therefore cladding splitting is a

problem long before 100% oxidation is reached. Another factor
influencing the oxidation rate would be the oxygen
concentration in the air in which the fuel is being stored.
Work by Blackburn et al[11] suggests that the oxidation rate
decreases by a factor of 2 in air depleted to 10% oxygen. This
factor, together with the decrease in the stored fuel heat
output due to fission product decay will decrease the rate of
fuel oxidation. Thus several factors including the fuel defect
size, temperature and oxygen concentration must be taken into
account in predicting potential fuel damage. What is
important, however, is to realize that some oxidation of UO_2
can be expected unless the temperatures are below 250°C.

5.0   SUMMARY AND CONCLUSIONS

        The primary goal of the Canadian Dry Storage Program
is to develop and demonstrate viable dry storage methods that
can be used as either complementary facilities to existing
water pools at reactor sites or in longer term (>50 yr)
centralized storage strategies. A dry storage concept, the
concrete canister, has been developed as a viable alternative
to building additional water pools. Irradiated fuel from
WNRE's research reactor is already being stored in concrete
canisters. A number of other dry storage concepts are being
assessed in Canada in order to establish the most promising
option for longer term irradiated fuel storage.

        Preliminary studies have shown that a single
multi-purpose concrete cask used for all phases of irradiated
fuel management: interim storage, transportation and disposal,
could result in substantial cost savings.

        A program is also underway to investigate the
durability of irradiated fuel in a controlled dry or moist
environment for long periods of time. The behaviour of
irradiated fuel stored in concrete canisters under seasonally
varying temperatures is also being monitored.

        Current available data indicates the mechanisms
leading to fuel degradation are complex. The temperature at
which fuel can be transported or stored safely in dry
environments will depend on several variables including:

        1)   fuel type;
        2)   fuel power history;
        3)   fuel condition: defected or undefected;
        4)   storage medium: air or inert-gas;
        5)   duration of transportation or storage.

        The definition of a maximum temperature for the dry
transportation or storage of irradiated fuel should therefore
be approached with caution. Further research is needed to
establish the existence of threshold temperatures for runaway
UO_2 oxidation and zircaloy clad stress corrosion cracking and
metal vapour embrittlement.

6.0 REFERENCES

[1]   S.J. Naqvi, B.P. Dalziel and R.C. Oberth. Irradiated
       Fuel Storage Program in Ontario Hydro. Paper
       presented to IAEA Advisory Group/Specialists Meeting
       on 'Spent Fuel Storage Alternatives.' Las Vegas,
       Nevada, USA. November 17-21, 1980.

[2]   C.E.L. Hunt, J.C. Wood, B.A. Surette and J.
       Freire-Canosa. 'Seventeen Years of Experience with
       Storage of Irradiated CANDU Fuel Under Water'. NACE
       Conference, Toronto, Canada. 1981.

[3]   B.P. Dalziel, S.J. Naqvi, and P.K.M. Rao. Long Term
       Storage of CANDU Irradiated Fuel, Ontario Hydro Report
       No. 82064, to be published.

[4]   M.M. Ohta. 'The Concrete Canister Program'. AECL
       Report No. 5965. February, 1978.

[5]   M.M. Ohta. 'Status of Dry Storage of Irradiated Fuel
       in Canada'. Paper presented to IAEA Advisory
       Group/Specialists Meeting on 'Spent Fuel Storage
       Alternatives'. Las Vegas, Nevada, USA. November
       17-21, 1980.

[6]   J. Freire-Canosa. 'A Preliminary Technical and
       Economic Evaluation of Concrete Casks for an
       Integrated Irradiated Fuel Management System'.
       Ontario Hydro Design and Development Report 82196,
       (1982).

[7]   K.M. Wasywich. Dry Storage Program - Technical Review
       and Status Report to December, 1981. AECL Report WNRE
       439-7, February, 1982.

[8]   J. Novak, I.M. Hastings. Private Communication,
       March, 1982.

[9]   S.B. Ainscough, and Oldfield, B.W. 'The Oxidation of
       Unirradiated Defected UO_2 Fuel Elements', United
       Kingdom Atomic Energy Authority, The Reactor Group,
       Reactor Fuel Element Laboratories, England, December,
       1964.

[10]   D.G. Boase and Vandergraff, T.T. 'The Canadian Spent
       Fuel Storage Canister: Some Material Aspects',
       Nuclear Technology, Volume 32, January, 1977.

[11]   P.E. Blackburn, J.E. Weishart, and E.A. Gulbranson.
       'Oxidation of Uranium Oxide', J. Phys. Chem., 62, 902
       (1958).

DISCUSSION

C. Melches, Spain

In your conclusions you mentioned the importance of the factor of the history of irradiated fuel.

Could you comment on your present thinking about the management of defective fuel ?

S.J. Naqvi, Canada

When I mentioned the fuel history as an important parameter in long term fuel behaviour, I was referring to fuel with no known defects but rather fuel which has been through power ramping and may have incipient cracks.

The fuel with known fuel defects e.g. pin holes, is canned within our pools and is stored separately. In dry storage, fuel with known defects can be handled similarly. However, we are studying long term fuel behaviour with fuel deliberately defected and with fuel which does not have any defect. We hope we can establish threshold conditions (e.g. temperature, air/inert gas environment) which would allow us to store fuel in dry air without undue concern for minor defects.

J. Fleisch, Federal Republic of Germany

Which parameters are foreseen to be continuously monitored when operating a commercial concrete cask facility in order to control e.g. leak-tightness.

S.J. Naqvi, Canada

We are hoping that our long term fuel behaviour experiments will establish that unlimited air storage of fuel in a commercial scale cask storage facility is safe. Hence we don't foresee the need for continuous monitoring for leak-tightness. However, if we do select inert gas as medium of storage then the usual leak-tightness monitoring would have to be done.

R. Kühnel, Federal Republic of Germany

Can the Canadian storage concept be reasonably applied in the case that the burn-up of HWR fuel increases considerably in the future ?

S.J. Naqvi, Canada

The answer is yes. We are looking into possible low enrichment-uranium fuel cycle (LEU) for CANDU's and thus higher burn-ups. However, even then the thermal power of our fuel bundles will be lower after 5 years waterpool cooling compared to say PWR fuel. So we don't see a problem since even for PWR fuel the US Silo program has shown encouraging results.

C.J. Ospina, Switzerland

Does the solar radiation build up the maximum heat load capacity of the CANDU-fuel storage concrete canisters, since they are continuously exposed, and if yes, how can you control the maximum fuel temperature during long term storage ?

S.J. Naqvi, Canada

When the fuel was loaded in our first canister containing easily retrievable fuel baskets (ERB), the temperature difference between the fuel and the outside ambient temperature was initially about 50°C. Since then the temperature difference has dropped gradually to about 25°C. If you want actual peak temperature reading during solar days, please see our report : K.W. Wasywich, Dry Storage Program - Technical Review and Status Report to December 1981 : AECL-Report WNRE 439-7, February 1982.

M. Trivino, Spain

Which standards or regulations are you considering for the testing of concrete casks to be used for transportation ?

S.J. Naqvi, Canada

We will attempt to meet revised IAEA requirements for transport casks.

M. Peehs, Federal Republic of Germany

You refer in your paper to incipient crack sizes of 40 % of wall thickness as a precondition for cracking the clad by SCC (Stress-Corrosion-Cracking) or liquid metal embrittlement. In my experience, incipient cracks of such size in the end-of-life condition are very unlikely to occur. If you have cracks of such size, due to ramping events in service, they will break the cladding during operation. My question is : what is your data base to assume such incipient cracks for storage conditions ?

S.J. Naqvi, Canada

I am afraid I do not have the data base here. However, I will attempt to find out the data base for you from CRNL (Chalk River Nuclear Laboratory).

THE ROLE OF DRY STORAGE AS A MEANS TO MEET
NATIONAL STORAGE NEEDS IN THE UNITED STATES

O. P. Gormley
U.S. Department of Energy
Washington, D.C., U.S.A.

ABSTRACT

This paper reviews the factors which seem to be governing the
application of dry storage in the United States. It suggests that
the rigorous technology is already well in hand and that it is
technical perceptions which will govern in the licensing process.
The paper traces the evolution of the U.S. program and shows how
nontechnical issues are involved and suggests that the technical
program managers should keep the perceptions in mind when defining
their objectives. Finally, the paper summarizes spent fuel
management cost data developed for the IAEA technical working
group on spent fuel management.

Dry storage is the most important alternative for meeting spent fuel storage needs in the United States.

Aside from technical considerations which can be adjusted to be roughly the same for all concepts, the degree to which any one spent fuel alternative will be used will depend on a number of factors:

o Need for storage space
o Certainty of availability--licensing and legislation
o Stability of Government policy which determines the help available from the Government
o "Public" acceptance on a national scale as influenced by anti-nuclear or what are called public interest groups
o Coincidence with the utility's policy and relationship with its local people
o Economics

My objective today is to present a different viewpoint. While we are doing very excellent work and have done so over the past several years in the technical and economic field, and we have heard all very good papers here dealing with those subjects, I would like to suggest that our work may have had the wrong focus. I would like to suggest that while we have been searching for technical and economic truth we may find that the solution, in fact, lies in administrative, political, and legislative remedies. These remedies are only very loosely dependent on technology and economics. After more than 5 years of vain attempts to provide Government spent fuel storage in the United States, we have found that the sophistication required of the technical and economic studies necessary to support decisions by utilities, lawmakers, public interest groups, and the Government is very low indeed. In fact, the decisions are generally based on perceptions and the perceptions are most successfully influenced by actual demonstrations.

In the case of water basin storage, the debate was won by a massive amount of evidence with respect to previous operations. In the case of dry storage, it is most likely that demonstration of abnormal conditions, including test to failure, will be needed because we do not have the time to amass the same overwhelming amount of data available for water basin storage. It seems that the testing and demonstration work done must be visibly related to the actual situation of storage under normal and abnormal conditions. Under these circumstances, it seems not to be so important to take statistically significant amounts of data under carefully controlled conditions as it is to let people see the tests and test conditions with their own eyes.

In spite of my support for and enthusiastic endorsement of the excellent paper presented by Mr. E. R. Johnson on the first day and also the information on relative costs for storage I will be presenting later myself, I consider storage costs to be a gross second order effect. To be sure, one should avoid serious mistakes like making a major investment in times of high interest rates in a facility with large initial investment costs. If the situation is also as uncertain as in the United States, it would seem that an investment in capacity for future years is to be avoided at all costs. But, is this really true? Our study of the value of a full core reserve /̲1̲/ estimated that each day a 1,000

1. "Full Core Reserve Study, A System Costing Analysis," The S. M. Stoller Corporation, November 28, 1971

MWe reactor would be shutdown, a utility would incur replacement power costs of $300,000 to $600,000. Based on a recent purchase by the U.S. Department of Energy, this would almost allow them to buy a dry storage cask every day capable of storing 10 MTU of spent fuel. When the utilities in the United States consider the new opportunities for mischief in the licensing process and the delays that go with them, optimization of storage costs within or among the various dry storage alternatives is of secondary importance. The primary concern is to reduce the risk of much greater financial losses which would be incurred if the actions by anti-nuclear groups or other intevenors in the licensing procedures delay the availability of dry storage beyond the point where the additional storage is needed. Thus, many utilities might build water basins if they had the time to do so. I do not believe any will though because many utilities do not have the time to build water basins and they will be forced to go to some dry storage alternative. In doing so, they may well prepare the way for the other utilities with later needs to solve their problems without resorting to a new water basin construction. On the other hand, the Government may provide emergency storage for only the earliest and most desperate needs. This will leave those utilities which now have time to build a water basin with a difficult decision to make. I believe they will wait to see what the future brings rather than taking action on water basin storage. What is clear though is that no utility is making a commitment to the use of dry storage unless all other options are foreclosed. Only one U.S. utility has indicated an intention to proceed and that indication is not firm.

If technical and storage cost considerations are not the primary ones, what are the primary considerations? One is need. Utility executives, legislators, and Government program managers are often willing to make correct decisions even if they are expensive and unpopular if there is a definite need.

For the past several years, we have been hampered in the United States with a constant and interrelated fluctuation of predicted storage capacity needs and proposed Government response to those needs. For example, as late as 1979 we were predicting that 70 metric tonnes of capacity was already needed; growing to 1,800 MTU in 1986 and almost 7,000 MTU in 1990. This resulted in a program which was large enough to gain considerable attention and attracted the interest of public interest groups. Partly as a result of their actions, the policy was changed to place the primary burden for spent fuel storage on the utilities but with the Government providing emergency capacity. This, some redefinitions of how we predicted requirements, and a growing concern on the part of the utilities that the Government would not provide the required storage capacity by the time it was needed led to a significant reduction in predicted requirements. By 1981 our prediction was that the first space would be needed in 1986 in the amount of 120 MTU, and that 1,800 MTU would be needed in 1990. Preliminary data available from a report to be published later this year shows that about 30 MTU would be required in 1985, a total of about 100 MTU in 1986, and roughly 1,260 MTU in 1990. The reduction in cumulative storage requirements has taken place as the result of utilities finding ways to put more fuel in their basins and more utilities making plans to ship fuel from one reactor to another.

The actions taken by some utilities to increase pool storage are quite extraordinary. For example, one utility is in the process of installing support posts in the area below the pool and then increasing the storage density only in those portions of the racks immediately above the posts. This allows them to complete an inspection which requires full core discharge, and they plan to have another reactor pool to ship fuel to before the extra storage is used up.

Actions like these have reinforced our perception that the utilities will go to great lengths to avoid introducing new technology like dry storage into the licensing process. Also, each time actions like this take place, the need for storage space decreases and, when it does, the resolve of national decision-makers to take some action in the near term weakens.

There are two developments on the horizon which will allow the utilities to put more fuel in their basins. The first is rod consolidation, which was described briefly by Mr. Johnson in his paper and is shown in the first two figures. We believe that this is a viable option and are supporting its demonstration.

FIGURE 1

FUEL DISASSEMBLY & CANNING FLOW PROCESS

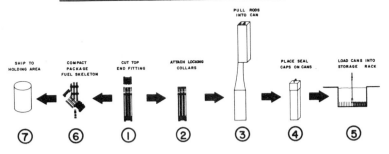

FIGURE 2

FUEL DISASSEMBLY AND
CANNING APPARATUS

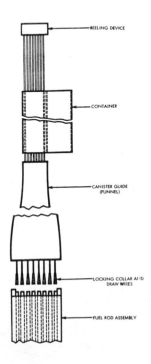

We had thought that rod consolidation would be the near-term
solution for utilities which have strong basins since you could
theoretically double the capacity of poison racks. However, we
believed that dry storage would be necessary for the early boiling
water reactor (BWR) reactors with storage pools fairly high in
containment. There is another development which we are cautiously
examining which may further reduce the use of dry storage. This
is shown in Figure 3.

FIGURE 3

**TYPICAL SPENT FUEL STORAGE
AND RACK INSTALLATION**

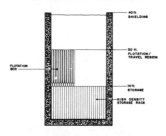

The actions taken by some utilities toincrease pool storage
are quite extraordinary. For example, one utility is in the
process of installing support posts in the area below the pool and
then increasing the storage density only in those portions of the
racks immediately above the posts. This allows them to complete
an inspection which requires full core discharge, and they plan to
have another reactor pool to ship fuel to before the extra storage
is used up.

Actions like these have reinforced our perception that the
utilities will go to great lengths to avoid introducing new
technology like dry storage into the licensing process. Also,
each time actions like this take place, the need for storage
space decreases and, when it does, the resolve of national
decisionmakers to take some action in the near term weakens.

There are two developments on the horizon which will allow the
utilities to put more fuel in their basins. The first is rod
consolidation, which was described briefly by Mr. Johnson in
his paper and is shown in the first two slides. We believe that
this is a viable option and are supporting its demonstration.

We had thought that rod consolidation would be the near-term
solution for utilities which have strong basins since you could
theoretically double the capacity of poison racks. However,
we believed that dry storage would be necessary for the early
boiling water reactor (BWR) reactors with storage pools fairly
high in containment. There is another development which we
are cautiously examining which may further reduce the use of
dry storage. This is shown in Figure 3.

The flotation boxes displace water and, thus, reduce the static
load on the pool. Their effect on seismic response is more
dramatic. Early testing work has suggested that a full complement
of float boxes flexibly attached to the top of the racks can
reduce the more limiting seismic loads by a factor of 40. If we
are successful in demonstrating this, it could reduce our fuel
storage needs even further.

The need to have a significant demonstration of the float box
concept brings me to the main point I am trying to make. In my
view, the primary consideration in the use of dry storage is the
question of certainty of availability. Right now that seems to be
a major deterrent which is preventing utilities from applying for
licenses. Our programs are designed to the best of our ability to
provide those demonstrations, and I will describe those later.
The U.S. Department of Energy can perform demonstrations of
technology but cannot obtain a license. Therefore, we have sought
cooperative agreements with utilities where we would provide
unlicensed demonstrations and other support to their initial use
of new storage techniques. We found little enthusiasm for this
because of our Nuclear Regulatory Commission procedures. The
problems lie with procedures established to give the public a
voice in the licensing process. When misused, they can result
in long delays while irrelevant arguments are made over whether
electricity is needed or not or whether the utility fully considered all its alternatives for dealing with spent fuel. We have
already talked about the cost of delays even excluding extensive
legal costs. As a result, late last year we and others turned our
attentions toward seeking legislative solutions to the real
impediments to the use of dry storage. As a result, we are now
seeing some legislation being considered by our Congress which
provides for generic licensing of storage casks. That is if one
utility gets a license for a dry cask, all the cask related safety
issues are excluded from discussion in further licensing actions
by other utilities. Certain of the frivolous but heatedly argued
environmental issues would also be excluded from debate. Best of
all, some safety valve storage capacity would be provided by the
Department for utilities which were trying to provide their own
storage space but which ran out of space during the argument over
the license. If this legislation passes, it will rank as first or
second on list of significant accomplishments by the public,
Government, and industry working together to reach a workable
compromise. The other contender for best accomplishment is that
seven commercial firms offered to build storage casks with 8 to 10
MTU capacity for $600,000 to $1 million.

The next most important step after achieving a reasonable and
reasonably certain licensing environment is to demonstrate that
the technology can be safely applied, thus obtaining favorable
perceptions with respect to safety and availability on the part of
the public, and utility and Government decisionmakers. We are
participants or expect to become participants in three dry storage
demonstrations.

The first is with the Tennessee Valley Authority where they will
apply for a license to store full fuel assemblies in two casks.
One is provided by GNS of the Federal Republic of Germany and the
other is provided by the Department of Energy from Ridihalgh,
Eggers, and Associates and Brooks and Perkins. If the generic
license provisions actually become law, the licensing of these two
types of casks at other utilities' sites will be greatly shortened.
Both casks will be loaded in 1983 and will contain instrumented
fuel to calibrate analytical techniques for multi-assembly casks.

Hopefully, some off normal experiments on restricted heat rejection will be conducted for further calibration. The second cooperative project is the shipment of consolidated rods to the Department of Energy in a regular transport cask also in 1983. These would be loaded into a storage cask and heat transfer and other tests conducted. Some of this fuel would end up in the Department's fuel performance archives as early life test samples. The third demonstration in 1984 would be the shipment of consolidated fuel in storage casks with protective devices added to allow the casks to meet the transportation accidents. At first this may seem a radical departure from past practices. However, some of the protective material for present shipping casks is already added to the containment package. This approach just goes a little further in that regard. Because of the extreme public sensitivity to transportation of radioactive materials we expect that the first shipments of this type will be made on an unlicensed basis by the Department of Energy.

I have only talked about metal casks. This is for two reasons. First, private enterprise and ingenuity have made this the cheapest of the options as reported by Mr. Johnson. Perhaps more importantly, those who make or influence decisions on which alternative to use prefer the cask. Utility decisionmakers prefer a metal cask because it can be added incrementally as an expense rather than as an investment, and they prefer it because it appears mobile and is out of the water table. They believe that the appearance of mobility is important to convince their public neighbors that the fuel will eventually be removed. Until we obtained our present perspective on economics, we carried development efforts on the drywell and, to a lesser extent, the concrete silo because of an intuitive belief that they should be less expensive than the metal casks. The public interest groups prefer the cask also because it appears mobile. They fear that drywells are less visible and attached to the ground so the fuel may stay there forever. They also argue that it is unsafe to store fuel in the ground water, which is close to the surface at all nuclear plants, no matter how many barriers are there.

The dry storage casks have another appeal to the public interest groups. Some of them seem to have real concerns over the safety of radioactive shipments. The prospect of gathering a great number of casks at a nuclear plant and shipping them out all at once on a single train appeals to them.

Once we have established proper perceptions of safety and availability, we need to have a strong technical program for licensing support and for those who want safe and effective operations whether they are required by licensing or not. We are in wonderful shape in this respect. We have all been working here for years and have done fine work. The only really important work to do is to calibrate the heat transfer analytical tools for multi-assembly casks and for packages of consolidated rods. After that we have only some adjustments to make. If the fuel is too hot, we can cool it longer in the pool or put less of it in the cask or we can put in an inert gas. While the problem of dealing with failed fuel at elevated temperatures in air is rather far down on the list of importance with respect to making dry storage a useful technique, it is probably the subject that makes me the most excited. If Dr. A. B. Johnson had not covered the subject so well yesterday morning, I would have had a lot to say about the $250\,^{\circ}C$ temperature limit and the need for data to support normal operations at 400°C and to assess abnormal operations at higher temperatures. I do have one thing to add to Dr. Johnson's information though. Early last week we agreed with the German Government to extensive collaboration in the dry fuel integrity area. I feel that there will be much mutual benefit.

In closing, this section of the paper, I want to go back over what I have attempted to do. While I firmly agree that rigorous technical work is necessary for safety and licensing and that economic studies are very important in deciding what business approach to take as a supplier to utilities or the U.S. Government, I have attempted to illustrate that they may only play a small role in determining whether dry storage will become a significant contributor to meeting spent fuel storage needs. In sharing this new found perspective with you, I am hoping you will join me in constantly searching for ways to accommodate the higher level concerns as we perform our technical work.

The objective of the next part of my talk is to introduce to you a paper prepared by the Expert Group on International Spent Fuel Management /̄2_/. This paper was part of an overall study of technical and economic considerations and is the joint effort of contributers from Austria, Switzerland, and the United States supplemented by additional data from Canada, the Federal Republic of Germany, Sweden, and the United Kingdom. The paper is available from the IAEA in two volumes and presents a good deal of detail on the designs of various storage concepts and some assumptions on operations.

The results are expressed as equations which are quite useful in developing a basis for understanding and comparing spent fuel storage costs.

The authors also expressed the results in terms of a fee that would be charged to fully recover all costs for providing the storage service. They selected a reference facility to compare the others against. The selected conditons were a water basin with 10,000 MTU capacity, a receiving period of 10 years at a rate of 1,000 MTU per year, a residence period of 10 years, and a 10-year removal period at 1,000 MTU per year. The resulting fee to be charged was about $100 per kg.

They then examined a number of variables asssociated with storage concepts, capacities, shipping distances, storage intervals, etc. The comparison of concepts showed the greatest impact ranging from a low of $70 per kg for the subsurface caisson or drywell to a high of about $200 per kg for the metal storage cask. There are good reasons for this difference being observed between the findings of this document and the findings presented by Mr. Johnson on the first day. First of all, his study was for about 1,000 MTU where as this study is for ten times that. As we shall see in a minute the impact of the facility capacity below

---

2.   "Subgroup A, Technical-Economic Considerations, Report on Task 3, Cost For Spent Fuel Management," August 1981

2,000 MTU can be dramatic. Also this study assumed all fuel to be encapsulated for all concepts which added considerably to relative costs for the sealed metal cask. Finally, the treatment of the discount rate was different which further emphasized the differences in approach. I believe each approach was correct for the purposes under consideration. Mr. Johnson's study was focused on the information needed by a utility faced with a near-term decision on what to do with excess fuel. It quite properly considered the different requirements for encapsulation and the interest payments for concepts with longer construction and utilization times. The IAEA study properly considered quantities more suitable for national fuel storage programs and compared concepts on as equal a basis as possible.

The next most important variable considered was the impact of variation in facility capacity. In fact, below 2,000 MTU this seems to be the effect which controls. For the concepts which require a large initial capital investment, such as the water basin and the vault, the fee does not come below $400 per kg until a capacity of 2,000 MTU is reached. However, the fee continues to drop all the way to 10,000 MTU where it reaches about $100 per kg. Casks and other truly modular concepts on the other hand start out with costs below $400 kg. However, the cask storage cost quickly levels out at $200 per kg.

The impact of other factors on the baseline concept is less noticeable. A five-fold increase in the discount rate translates to less than a 20 percent increase in cost. Adding aircraft crash protection increases the cost 30 percent. A five-fold increase in transportation distance adds about 25 percent to the fee. Tripling train speed on a 2,000 mile journey saves less than 10 percent, and increasing the storage duration in a situation where the increase occurs 25 years in the future costs almost nothing at all even if the effective discount rate is only 2 percent.

DISCUSSION

<u>P. Erlenwein</u>, Federal Republic of Germany

As dry storage is a rather new technology, do you feel that it may be helpful to carry out risk assessment studies for this technology and to use the results as criteria for decision making and to increase public acceptance ?

<u>O.P. Gormley</u>, United States

Risk assessments will be required to satisfy the requirement for a complete record in the licensing process. I personally believe that the public legislators are not impressed by risk assessments. They seem to prefer hardware demonstrations.

<u>E.O. Maxwell</u>, United Kingdom

You say that in your view the primary consideration in the use of dry storage is the question of certainty of availability and that seems to be a major deterrent which is preventing utilities from applying for licences. But is it not true that a safety case for vault type storage is available (which could be built in about 3 years) but the "deterrent" really is in the ramifications and delays caused by public opinion and your licensing authority.

<u>O.P. Gormley</u>, United States

Yes. The entire problem is the perception by utilities that their implementation of anything other than the types of water basin storage already licensed in the United States will be delayed by lengthy intervention in the licensing process. They fear that the licensing process will drag on beyond the time when they need the storage.

<u>J.R. Haddon</u>, United Kingdom

Did the various offers for supply of storage flasks guarantee licensability of the flask in each case ?

<u>O.P. Gormley</u>, United States

Clearly a storage flask is of no value if it is not licensable. In the case of DOE storage casks a licence would not be necessary and therefore licensability was not an issue. However, the demonstration was designed to show that licensing requirements could be met and the supplier was obliged to provide a safety analysis and other comprehensive data for this purpose.

<u>J. Fleisch</u>, Federal Republic of Germany

You mentioned NRC saying that the inert gas atmosphere must be controlled or guaranteed when storing fuel at temperatures above 250°C whereas even for storage is unlimited air no problems are expected. Was that request directed towards a specific dry storage system or was it a general requirement ? Because in the case of the German metal casks, helium gas atmosphere could be guaranteed and continuously checked by the interspace monitoring in combination with the pressure between the two lid barriers.

O.P. Gormley, United States

        The request is directed to all dry storage systems. I would
think they could find a way that they could question whether the
indication from between the lids was representative of the storage
environment.

B.R. Teer, United States

        The US has a one step licensing process for on-site spent
fuel storage whereas Germany issues a construction permit prior to
an operating licence. Do you think this is a significant deterrent
to getting a US utility to apply for a licence ?

O.P. Gormley, United States

        Yes. However, if the generic licensing provisions pass our
Congress this will turn out to be an advantage. Right now it is a
disadvantage because a utility which has three years or less to get
a licence has to wait two years to find out whether its first option
will be accepted by NRC.

B.R. Teer, United States

        Do you think the US licensing requirements discriminate
against dry storage techniques in favour of water basins ?

O.P. Gormley, United States

        Definitely. Dry storage containment requirements and fuel
performance requirements are much more stringent for dry storage than
for water storage. This is another area which requires non-technical
attention.

G.A. Brown, United Kingdom

        I found it refreshing to hear you say that : "I consider
storage costs to be a gross second order effect".

        When the chips are down and you are actually placing orders,
how do you propose defending that you may not be going for the
cheapest option ?

O.P. Gormley, United States

        It is really the utilities which must defend their actions.
I think the utility rate regulators will agree with them that the
risk of shutdown costs for a nuclear plant is so great that they are
justified in selecting an option which may be more expensive but
which offers greater flexibility during these uncertain times and
which is viewed more favourably by the public and licensing bodies.

P.G.K. Doroszlay, Switzerland

        You emphasized the importance of mobility of storage casks.
The REA-Cask needs an overpack for transportation which is even not
yet designed. There are also doubts concerning the licensability of
nodular cast iron for transport in the United States. Do you think,
the REA or the cast iron casks will ever be licensed for transport ?

O.P. Gormley, United States

        Yes I do. At least two US firms are considering overpacks
for a storage cask and at least one is actively designing one. It
may be that some sort of overpack or even heating would be required
for the cast iron casks but I believe that shipment in those casks
will be possible as well.

R. Kühnel, Federal Republic of Germany

        What is the necessary age of fuel to be consolidated in
respect to temperature limits in the (wet) storage pool ?

O.P. Gormley, United States

        I do not really know. I am not aware of any problem with
aged fuel.

R. Kühnel, Federal Republic of Germany

        Is the US public aware of the fact that the utilities choose
a more expensive -(not safer)- option in order to overcome the public
acceptance problem, which the public has to pay for ?

O.P. Gormley, United States

        I believe they are. Actually the difference is rather small.
Also they keep electing people to the Government who want the more
expensive measures taken.

A.H.M. Janssen, The Netherlands

        Is an environmental impact statement required in the United
States with respect to interim storage ?

E.R. Johnson, United States

        Yes.

E.O. Maxwell, United Kingdom

        Do you not have a criticality problem when using the rod
consolidation system in your pools without the use of boronated
racks ?

O.P. Gormley, United States

        No. As the rods are brought closer together the metal to
water ratio increases to a less reactive condition. Critical
assembly experiments confirmed this.

H.J. Wervers, The Netherlands

        With respect to the use of floaters to reduce water load on
bottom of pool, I am puzzled as to how the effect of radiation
beaming through air-filled floaters is tackled ?

O.P. Gormley, United States

This is not a problem. First the radioactivity of older fuel is much less. Perhaps more importantly the floaters are most useful in pools where the fuel is handled over the top of the racks. Therefore, there is 15 ft extra water depth for fuel handling purposes only.

B. Vriesema, The Netherlands

Would Mr. Tabe comment on the difference in cost/kg U given by him and e.g. by Mr. Johnson ?

E.R. Johnson, United States

The casks were designed differently; the Canadian design was a smaller sealed cask, while the JAI (Johnson Associated, Inc.) design was a natural convection cooled cask. The Canadian design was for lower burnup CANDU fuel, while the JAI design was for 33,000 MWD/MTU* burnup PWR fuel. The Canadian design accommodated about 4 MTU* of fuel in the sealed silo, whereas the JAI design accommodated 4-PWR assemblies or about 1.8 MTU* of fuel. The JAI design silo was estimated to cost about $ 75,000 ($ 58,000 for the silo and $ 17,000 for the inner liner assembly). The JAI design was clearly more complex (by necessity) than the Canadian one. Moreover, the JAI design included concrete to a specification of high compressive strength and had heavy rebar imbedded in the concrete.

The JAI estimates included all commercial costs (estimates) that were believed to be applicable, whereas Mr. Tabe indicated that all such costs were not included.

A detailed comparison of the cost estimates has not been made as yet.

---

* MTU = Metric Tonnes Uranium.

PANEL DISCUSSION

Chairman - Président

H.R. KONVICKA
(Austria)

TABLE RONDE

A.B. JOHNSON Jr., United States

B. LOPEZ PEREZ, Spain

E.O. MAXWELL, United Kingdom

M. PEEHS, Federal Republic of Germany

W.T. POTTER, OECD Nuclear Energy Agency

PANEL   TABLE RONDE

The panel was convened to try to extract the essential
conclusions to be drawn from the reports presented during the course
of the workshop; to discuss future trends in dry storage techniques
and their role in meeting the interim storage needs of countries with
different nuclear programmes; and to consider relevant research and
development work.

The main points arising from the Panel's discussions may be
summarized as follows :

1.   It was felt that dry storage could now be considered to be
a proven technology. Within the last few years, various dry storage
concepts have been successfully demonstrated in large scale
experiments with spent fuel representative of today's generation of
LWR and HWR fuels. The predictions made for these experiments have
been confirmed by the experimental results and proved to be very
conservative. Operators have up to 18 years' experience with dry
storage of fuel from other types of reactors, such as gas-cooled (GCR),
high temperature (HTR) and fast breeder (FBR).

2.   Available evidence shows that this method of storage is not
likely to lead to unacceptable degradation of the fuel cladding or
result in any other damage to the fuel. However, research and
development work is being continued to further reinforce this
evidence. This includes studies on oxidization at varying temperatures
and humidities; on the behaviour of defective fuel; on waterlogging;
on potential stress corrosion cracking mechanisms and chemical agents,
and on stress rupture. For the future, special emphasis will be
placed on investigations employing high burn-up spent fuel.

3.   In several countries, licence applications for dry storage
have been submitted to the relevant licensing bodies. Since there
exists a sound supporting technical data base, no major difficulties
are expected. For industrial-scale GCR fuel storage, the licensing
procedure may be considered as established.

4.   Rapid development of various dry storage concepts lead to
some confusion in nomenclature. It was felt that the nomenclature
suggested for dry storage concepts by the IAEA Expert Group on
International Spent Fuel Management should be used to the maximum
extent possible.

This nomenclature differentiates :

- caisson storage (strong protective substitute overpack, not movable);

- cask storage (heavy, strong building-like construction, different heat removal mechanisms :
  - active cooling (forced convection), either directly or indirectly (heat removal medium in direct or indirect contact with stored fuel assembly);
  - passive cooling (natural convection), either directly or indirectly.

5.   Whilst wet storage for spent fuel is indispensable for a period up to 12 months after discharge of the spent fuel from the reactor to allow for radioactive decay and initial cooling, the introduction of dry storage techniques very much depends on the needs of a country's nuclear programme. Dry storage appears to offer technical advantages such as the elimination of possible loss of coolant accidents and the significant reduction of secondary generated radioactive waste. Economic advantages have to be identified on a case-by-case basis, since numerous factors have to be considered such as the storage concept, the size of the facility (amount of spent fuel which has to be accommodated), the location of the facility (either at the reactor site or at an independent storage site), and the time scale (availability of future steps envisaged in the fuel cycle). Thus, no universally applicable solution for spent fuel management by dry storage can be offered.

6.   The question of public attitudes towards nuclear energy was also touched upon. It was acknowledged that at present, good storage practice was not of itself likely to enhance public acceptance; on the other hand, however, evidence of unsafe storage methods could be extremely damaging to public confidence. Therefore, every effort must be made to improve the outstanding safety record of nuclear energy even further, particularly in regard to spent fuel storage practices.

LIST OF PARTICIPANTS
LISTE DES PARTICIPANTS

AUSTRIA - AUTRICHE

KONVICKA, H.R., Consultant OECD/NEA, Austrian Research Centre
    Seibersdorf (ÖFZS), Lenaugasse 10, A-1082 Vienna

BELGIUM - BELGIQUE

PAULUIS, G., SYNATOM, 13, Avenue Marnix, B-1050 Bruxelles

CANADA

NAQVI, S.J., Ontario Hydro, Nuclear Materials Management
    Department, 700 University Avenue, Toronto, Ontario M5G 1X6

NOVAK, Y., Ontario Hydro, 700 University Avenue, Toronto,
    Ontario M5G 1X6

TABE, T., Atomic Energy of Canada Limited, Whiteshell Nuclear
    Research Establishment, Pinawa, Manitoba ROE 1L0

FINLAND - FINLANDE

KANGAS, J., Teollisuuden Voima Oy/TVO Power Company,
    Kutojantie 8, SF-02630 Espoo 63

TAKALA, H.J., Inspector, Institute of Radiation Protection,
    Department of Reactor Safety, P.O. Box 268,
    SF-00101 Helsinki 10

FRANCE

GEFFROY, J., Chef de Laboratoire, Commissariat à l'Energie
    Atomique, 91191 Gif-sur-Yvette

LANGLOIS, J.P., Ingénieur au DIP, Commissariat à l'Energie
    Atomique - DgAEN, 29-33, rue de la Fédération, B.P. 510,
    75752 Paris Cedex 15

SUGIER, A., Assistante du Directeur des Applications
Energétiques
    Nucléaires, Commissariat à l'Energie Atomique - DgAEN,
    B.P. 510, 75752 Paris Cedex 15

VERDANT, R., Ingénieur au Département des Programmes,
    Commissariat à l'Energie Atomique, 29-33, rue de la
    Fédération, B.P. 510, 75752 Paris Cedex 15

FEDERAL REPUBLIC OF GERMANY - REPUBLIQUE FEDERALE D'ALLEMAGNE

BOKELMANN, R.F., Kraftwerk Union AG, Postfach 962
    D-6050 Offenbach

CHRIST, R., Transnuklear GmbH, Postfach 110030, D-6450 Hanau 11

ERLENWEIN, P., Gesellschaft für Reaktorsicherheit,
    Glockengasse/2, D-5000 Köln 1

FLEISCH, J., DWK, Hamburger Allee 2-4, D-3000 Hannover 1

Von HEESEN, W., STEAG Kernenergie GmbH, Bismarckstrasse 54,
    D-4300 Essen

KEESE, H., General Manager, Transnuklear GmbH, Postfach 110030,
    D-6450 Hanau 11

KIOES, S., NUKEM GmbH, Postfach 110080, D-6450 Hanau 11

KUHNEL, R., Kraftwerk Union AG, Wiesenstrasse 35,
    D-4330 Mülheim an der Ruhr

PEEHS, M., Kraftwerk Union AG, Hammerbacherstrasse 12-14,
    D-8520 Erlangen

PORSCH, G., NUKEM GmbH, Postfach 110080, D-6450 Hanau 11

RAMCKE, K., Preussische Elektrizitäts-Aktiengesellschaft
    Tresckowstrasse 5, D-3000 Hannover 91

SCHONFELD, R., NUKEM GmbH, Postfach 110080, D-6450 Hanau 11

STAHL, D., NUKEM GmbH, Postfach 110080, D-6450 Hanau 11

TIMME, J., NUKEM GmbH, Postfach 110080, D-6450 Hanau 11

ITALY - ITALIE

CUZZANITI, M., ENEL - DCO, Via G.B. Martini 3, I-00198 Roma

JAPAN - JAPON

ADACHI, M.A., Tokai Research Establishment, Tokai-Mura,
    Naka-gun, Ibaraki-ken

THE NETHERLANDS - PAYS-BAS

CODEE, H.D.K., Ministry of Health and Environmental Protection,
    P.O. Box 5811, 2280 HV Ryswyk

EENDEBAK, B.Th., N.V. KEMA, Utrechtseweg 310, P.O. Box 9035,
    6800 ET Arnhem

JANSSEN, A.H.M., Ministry of Economic Affairs, Directie
    Elektriciteit en Kernenergie, Post Box 20101, The Hague

VRIESEMA, B., Netherlands Energy Research Foundation (ECN),
    P.O. Box 1, 1755 LE Petten

WERVERS, H.J., Netherlands Energy Research Foundation (ECN),
    P.O. Box 1, 1755 LE Petten

SPAIN - ESPAGNE

ARROYO RUIPEREZ, J., Junta de Energia Nuclear, Avenida
    Complutense s/n, Ciudad Universataria, Madrid 3

BARCENA, P., ENUSA, Santiago Rusinol 12, Madrid 3

CALLEJA MAESTRE, J.M., Central Nuclear Valdecaballeros,
    Claudio Coello 53, Madrid 1

CARO MANSO, R., Junta de Energia Nuclear, Avenida
    Complutense s/n, Ciudad Universitaria, Madrid 3

CHAMERO FERRER, A., Junta de Energia Nuclear, Avenida
    Complutense s/n, Ciudad Universitaria, Madrid 3

DE MATIAS, E., ENUSA, Santiago Rusinol 12, Madrid 3

ESTEBAN NAUDIN, A., Junta de Energia Nuclear, Avenida
    Complutense s/n, Ciudad Universitaria, Madrid 3

FERNANDEZ SANCHEZ, T., Union Electrica S.A., Capitan Haya 53,
    Madrid 20

FORNEZ SANCHEZ, A., Junta de Energia Nuclear, Avenida
    Complutense s/n, Ciudad Universitaria, Madrid 3

GARCIA RAMIREZ, M., ENUSA, Santiago Rusinol 12, Madrid 3

GISPERT BENACH, M., Junta de Energia Nuclear, Avenida
    Complutense s/n, Ciudad Universitaria, Madrid 3

GONZALEZ GOMES, J.L., NUKEM GmbH, Sucursal en Espana,
    Hermosilla 57, Madrid 1

GONZALEZ DE UBIETA, A., Unidad Electrica S.A. (UNESA),
    Francisco Gervas 3, Madrid 20

GUZMAN MATAIX, J.L., Almaraz Nuclear Power Plant, Princesa 3,
    Madrid 8

HERRANZ, R., Assistant Director to the General Manager,
    ENUSA, Santiago Rusinol 12, Madrid 3

LATOVA TRIGO, J.J., Equipos Nucleares S.A. (ENSA), Apartade 304,
    Santander

LEON, J.R., Almaraz Nuclear Power Plant, Princesa 3, Madrid 8

LOPEZ PEREZ, B., Director of the Back End of the Nuclear Fuel
    Cycle, Junta de Energia Nuclear, Avenida Complutense s/n,
    Ciudad Universitaria, Madrid 3

MOCHON, J.L., Physicist, Hidroelectrica Espanola S.A.,
    Hermossilla 3, Madrid 1

MELCHES SERRANO, C., ENUSA, Santiago Rusinol 12, Madrid 3

MONTES PONCE DE LEON, M., Junta de Energia Nuclear, Avenida
    Complutense s/n, Ciudad Universitaria, Madrid 3

OTERO DE LA GANDARA, J.L., Junta de Energia Nuclear, Avenida
    Complutense s/n, Ciudad Universitaria, Madrid 3

OYARZABAL, I., ENUSA, Santiago Rusinol 12, Madrid 3

PENA GUTIERREZ, J., Junta de Energia Nuclear, Avenida
    Complutense s/n, Ciudad Universitaria, Madrid 3

PERLADO MARTIN, M., Junta de Energia Nuclear, Avenida
    Complutense s/n, Madrid 3

POVAR ALBILLOS, G., Equipos Nucleares S.A. (ENSA), Apartado 304,
    Santander

PUGA, J., ENUSA, Santiago Rusinol 12, Madrid 3

RAMIREZ ONTALBA, E., ENUSA, Division Tratamiento del Combustible
    Irradiado, Santiago Rusinol 12, Madrid 3

RAMOS SALVADOR, L., Jefe Seccion de Tratamiento de Combustibles
    Irradiados, Junta de Energia Nuclear, Avenida Complutense s/n
    Ciudad Universitaria, Madrid 3

RODRIGUEZ PARRA, M., Jefe Proyecto Almacenamiento de
    Combustibles Irradiados, Junta de Energia Nuclear, Avenida
    Complutense s/n, Ciudad Universitaria, Madrid 3

RODRIGUEZ-SOLANO Y PASTRANA, R., Junta de Energia Nuclear,
    Avenida Complutense s/n, Ciudad Universitaria, Madrid 3

ROVIRA MARTI, A., Junta de Energia Nuclear, Avenida
    Complutense s/n, Ciudad Universitaria, Madrid 3

TRIVINO, M., Equipos Nucleares S.A. (ENSA), Apartado 304,
    Santander

URIARTE, A., Irradiated Fuels Division Chief, Junta de Energia
    Nuclear, Avenida Complutense s/n, Ciudad Universitaria
    Madrid 3

VERON, P., Equipos Nucleares S.A. (ENSA), Apartado 304,
    Santander

DOROSZLAI, P.G.K., Electrowatt Engineering Services,
    CH-8022 Zürich

GUT, W., Notdostschweizerische Kraftwerke AG, Parkstrasse 23,
    CH-5401 Baden

HOOP, F.J., Elektrizitäts-Gesellschaft Laufenburg AG,
    Dufourstrasse 101, Postfach, CH-8022 Zürich

MEYER, L., Electrowatt Engineering Services,
    CH-8022 Zürich

OSPINA, C.J., Swiss Federal Institute for Reactor Research
    (EIR), CH-5303 Würenlingen

TAORMINA, A.M., Nuclear Assurance Corporation, Weinberg-
    Strasse 9, CH-8001 Zürich

UNITED KINGDOM - ROYAUME-UNI

BRADLEY, N., National Nuclear Corporation Ltd., Warrington Road,
    Risley, Warrington WA3 6BZ

BROWN, G.A., Central Electricity Generating Board, Generation
    Development and Construction Division, Barnett Way,
    Barnwood, Gloucester GL4 7RS

HADDON, J.R., Overseas Sales Executive, GEC Energy Systems Ltd.
    Cambridge Road, Whetstone, Leicester LE8 3LH

MAXWELL, E.O., Fuelling Development Manager, Central
    Electricity Generating Board, North Western Region, Birdhall
    Road, Cheadle Heath, Stockport, Cheshire SK3 0XA

PEARCE, R.J., Central Electricity Generating Board, Berkeley
    Nuclear Laboratories, Berkeley, Gloucestershire

PRICE, M.S.T., United Kingdom Atomic Energy Authority,
    AEE Winfrith, Dorchester, Dorset DT2 8DH

RAWSON, D.F., Engineer, Process Technology and Safety - E433
    United Kingdon Atomic Energy Authority, Risley,
    Warrington, Cheshire

TANNER, M.C., British Nuclear Fuels Limited, Room A202, Risley, Warrington, Cheshire WA3 6AS

WHEELER, D.J., Design Consultant, GEC - Energy Systems Ltd., Cambridge Road, Whetstone, Leicester LE8 3LH

UNITED STATES - ETATS-UNIS

GORMLEY, O.P., Director, Division of Waste Treatment and Spent Fuel Management, U.S. Department of Energy, NE-340, Washington DC 20545

JOHNSON, Jr., A.B., Senior Staff Scientist, Battelle-Northwest, 3720 Bldg., 300 Area, Box 999, Richland, Washington

JOHNSON, E.R., President, E.R. Johnson Associates Inc., 11702 Bowman Green Drive, Reston, Virginia 22090

TEER, B.R., Vice President, Transnuclear Inc., One North Broadway, White Plains, New York 10601

INTERNATIONAL ATOMIC ENERGY AGENCY

AGENCE INTERNATIONALE POUR L'ENERGIE ATOMIQUE

GALKIN, V., I.A.E.A., Wagramerstrasse 5, P.O. Box 100, A-1400 Vienna, (Austria)

OECD NUCLEAR ENERGY AGENCY

AGENCE DE L'OCDE POUR L'ENERGIE NUCLEAIRE

POTTER, W.T., OECD/NEA, Nuclear Development Division, 38, Boulevard Suchet, 75016 Paris, (France)

QUARMEAU, S., OECD/NEA, Nuclear Development Division, 38, Boulevard Suchet, 75016 Paris, (France)

SOME NEW PUBLICATIONS OF NEA

ACTIVITY REPORTS

Activity Reports of the OECD Nuclear Energy Agency (NEA)
— 9th Activity Report (1980)
— 10th Activity Report (1981)

Free on request — Gratuits sur demande

Annual Reports of the OECD HALDEN Reactor Project
— 19th Annual Report (1978)
— 20th Annual Report (1979)

Free on request — Gratuits sur demande

● ● ●

INFORMATION BROCHURES

— OECD Nuclear Energy Agency: Functions and Main Activities

— NEA at a Glance
International Co-operation for Safe Nuclear Power
The NEA Data Bank

Free on request — Gratuits sur demande

● ● ●

NUCLEAR ENERGY PROSPECTS

Nuclear Energy Prospects to 2000
(A joint Report by NEA/IEA)

£7.00   US$14.00   F70,00

QUELQUES NOUVELLES PUBLICATIONS DE L'AEN

RAPPORTS D'ACTIVITÉ

Rapports d'activité de l'Agence de l'OCDE pour l'Énergie Nucléaire (AEN)
— 9e Rapport d'Activité (1980)
— 10e Rapport d'Activité (1981)

Rapports annuels du Projet OCDE de réacteur de HALDEN
— 19e Rapport annuel (1978)
— 20e Rapport annuel (1979)

BROCHURES D'INFORMATION

— Agence de l'OCDE pour l'Énergie Nucléaire : Rôle et principales activités
— Coup d'œil sur l'AEN
— Une coopération internationale pour une énergie nucléaire sûre
— La Banque de Données de l'AEN

PERSPECTIVES DE L'ÉNERGIE NUCLÉAIRE

Perspectives de l'Énergie Nucléaire jusqu'en 2000
(Rapport conjoint AEN/AIE)

SCIENTIFIC AND TECHNICAL PUBLICATIONS

PUBLICATIONS SCIENTIFIQUES ET TECHNIQUES

NUCLEAR FUEL CYCLE

World Uranium Potential —
An International Evaluation (1978)

£7.80   US$16.00   F64.00

Uranium — Ressources, Production and Demand (1982)

£9.90   US$22.00   F99,00

Nuclear Energy and Its Fuel Cycle: Prospects to 2025

£11.00   US$24.00   F110,00

Uranium Exploration Methods — Review of the NEA/IAEA R & D Programme
(Proceedings of the Paris Symposium, 1982) [in preparation].

£24.00   US$48.00   F240,00

LE CYCLE DU COMBUSTIBLE NUCLÉAIRE

Potentiel mondial en uranium —
Une évaluation internationale (1978)

Uranium — ressources, production et demande (1982)

L'énergie nucléaire et son cycle de combustible : perspectives jusqu'en 2025

Les méthodes de prospection de l'uranium — Examen du programme AEN/AIEA de R & D
(Compte rendu du symposium de Paris, 1982) [en préparation].

● ● ●

SCIENTIFIC INFORMATION

Calculation of 3-Dimensional Rating Distributions in Operating Reactors
(Proceedings of the Paris Specialists' Meeting, 1979)

£9.60   US$21.50   F86.00

Nuclear Data and Benchmarks for Reactor Shielding
(Proceedings of a Specialists' Meeting, Paris, 1980)

£9.60   US$24.00   F96,00

INFORMATION SCIENTIFIQUE

Calcul des distributions tridimensionnelles de densité de puissance dans les réacteurs en cours d'exploitation (Compte rendu de la Réunion de spécialistes de Paris, 1979)

Données nucléaires et expériences repères en matière de protection des réacteurs (Compte rendu d'une réunion de spécialistes, Paris, 1980)

● ● ●

RADIATION PROTECTION

RADIOPROTECTION

Iodine-129
(Proceedings of an NEA Specialist Meeting, Paris, 1977)

Iode-129
(Compte rendu d'une réunion de spécialistes de l'AEN, Paris, 1977)

£3.40   US$7.00   F28,00

Recommendations for Ionization Chamber Smoke Detectors in Implementation of Radiation Protection Standards (1977)

Recommandations relatives aux détecteurs de fumée à chambre d'ionisation en application des normes de radioprotection (1977)

Free on request — Gratuit sur demande

Radon Monitoring
(Proceedings of the NEA Specialist Meeting, Paris, 1978)

Surveillance du radon
(Compte rendu d'une réunion de spécialistes de l'AEN, Paris, 1978)

£8.00   US$16.50   F66,00

Management, Stabilisation and Environmental Impact of Uranium Mill Tailings (Proceedings of the Albuquerque Seminar, United States, 1978)

Gestion, stabilisation et incidence sur l'environnement des résidus de traitement de l'uranium
(Compte rendu du Séminaire d'Albuquerque, États-Unis, 1978)

£9.80   US$20.00   F80,00

Exposure to Radiation from the Natural Radioactivity in Building Materials (Report by an NEA Group of Experts, 1979)

Exposition aux rayonnements due à la radioactivité naturelle des matériaux de construction
(Rapport établi par un Groupe d'experts de l'AEN, 1979)

Free on request — Gratuit sur demande

Marine Radioecology
(Proceedings of the Tokyo Seminar, 1979)

Radioécologie marine
(Compte rendu du Colloque de Tokyo, 1979)

£9.60   US$21.50   F86.00

Radiological Significance and Management of Tritium, Carbon-14, Krypton-85 and Iodine-129 arising from the Nuclear Fuel Cycle (Report by an NEA Group of Experts, 1980)

Importance radiologique et gestion des radionucléides : tritium, carbone-14, krypton-85 et iode-129, produits au cours du cycle du combustible nucléaire
(Rapport établi par un Groupe d'experts de l'AEN, 1980)

£8.40   US$19.00   F76,00

The Environmental and Biological Behaviour of Plutonium and Some Other Transuranium Elements (Report by an NEA Group of Experts, 1981)

Le comportement mésologique et biologique du plutonium et de certains autres éléments transuraniens (Rapport établi par un Groupe d'experts de l'AEN, 1981)

£4.60   US$10.00   F46,00

Uranium Mill Tailings Management (Proceedings of two Workshops)

La gestion des résidus de traitement de l'uranium
(Compte rendu de deux réunions de travail)

£7.20   US$16.00   F72,00

RADIOACTIVE WASTE MANAGEMENT

GESTION DES DÉCHETS RADIOACTIFS

Objectives, Concepts and Strategies for the Management of Radioactive Waste Arising from Nuclear Power Programmes (Report by an NEA Group of Experts, 1977)

Objectifs, concepts et stratégies en matière de gestion des déchets radioactifs résultant des programmes nucléaires de puissance (Rapport établi par un Groupe d'experts de l'AEN, 1977)

£8.50   US$17.50   F70,00

Treatment, Conditioning and Storage of Solid Alpha-Bearing Waste and Cladding Hulls (Proceedings of the NEA/IAEA Technical Seminar, Paris, 1977)

Traitement, conditionnement et stockage des déchets solides alpha et des coques de dégainage (Compte rendu du Séminaire technique AEN/AIEA, Paris, 1977)

£7.30   US$15.00   F60,00

Storage of Spent Fuel Elements (Proceedings of the Madrid Seminar, 1978)

Stockage des éléments combustibles irradiés (Compte rendu du Séminaire de Madrid, 1978)

£7.30   US$15.00   F60,00

In Situ Heating Experiments in Geological Formations (Proceedings of the Ludvika Seminar, Sweden, 1978)

Expériences de dégagement de chaleur in situ dans les formations géologiques (Compte rendu du Séminaire de Ludvika, Suède, 1978)

£8.00   US$16.50   F66,00

Migration of Long-lived Radionuclides in the Geosphere (Proceedings of the Brussels Workshop, 1979)

Migration des radionucléides à vie longue dans la géosphère (Compte rendu de la réunion de travail de Bruxelles, 1979)

£8.30   US$17.00   F68,00

Low-Flow, Low-Permeability Measurements in Largely Impermeable Rocks (Proceedings of the Paris Workshop, 1979)

Mesures des faibles écoulements et des faibles perméabilités dans des roches relativement imperméables (Compte rendu de la réunion de travail de Paris, 1979)

£7.80   US$16.00   F64,00

On-Site Management of Power Reactor Wastes (Proceedings of the Zurich Symposium, 1979)

Gestion des déchets en provenance des réacteurs de puissance sur le site de la centrale (Compte rendu du Colloque de Zurich, 1979)

£11.00   US$22.50   F90,00

Recommended Operational Procedures for Sea Dumping of Radioactive Waste (1979)

Recommandations relatives aux procédures d'exécution des opérations d'immersion de déchets radioactifs en mer (1979)

Free on request — Gratuit sur demande

Guidelines for Sea Dumping Packages of Radioactive Waste (Revised version, 1979)

Guide relatif aux conteneurs de déchets radioactifs destinés au rejet en mer (Version révisée, 1979)

Free on request — Gratuit sur demande

Use of Argillaceous Materials for the Isolation of Radioactive Waste (Proceedings of the Paris Workshop, 1979)

Utilisation des matériaux argileux pour l'isolement des déchets radioactifs (Compte rendu de la Réunion de travail de Paris, 1979)

£7.60   US$17.00   F68,00

Review of the Continued Suitability of the Dumping Site for Radioactive Waste in the North-East Atlantic (1980)

Réévaluation de la validité du site d'immersion de déchets radioactifs dans la région nord-est de l'Atlantique (1980)

Free on request — Gratuit sur demande

Decommissioning Requirements in the Design of Nuclear Facilities (Proceedings of the NEA Specialist Meeting, Paris, 1980)

Déclassement des installations nucléaires : exigences à prendre en compte au stade de la conception (Compte rendu d'une réunion de spécialistes de l'AEN, Paris, 1980)

£7.80   US$17.50   F70,00

Borehole and Shaft Plugging (Proceedings of the Columbus Workshop, United States, 1980)

Colmatage des forages et des puits (Compte rendu de la réunion de travail de Columbus, États-Unis, 1980)

£12.00   US$30.00   F120,00

Radionucleide Release Scenarios for Geologic Repositories (Proceedings of the Paris Workshop, 1980)

Scénarios de libération des radionucléides à partir de dépôts situés dans les formations géologiques (Compte rendu de la réunion de travail de Paris, 1980)

£6.00   US$15.00   F60,00

Research and Environmental Surveillance Programme Related to Sea Disposal of Radioactive Waste (1981)

Programme de recherches et de surveillance du milieu lié à l'immersion de déchets radioactifs en mer (1981)

Free on request — Gratuit sur demande

Cutting Techniques as related to Decommissioning of Nuclear Facilities (Report by an NEA Group of Experts, 1981)

Techniques de découpe utilisées au cours du déclassement d'installations nucléaires (Rapport établi par un Groupe d'experts de l'AEN, 1981)

£3.00   US$7.50   F30.00

Decontamination Methods as related to Decommissioning of Nuclear Facilities (Report by an NEA Group of Experts, 1981)

Méthodes de décontamination relatives au déclassement des installations nucléaires (Rapport établi par un Groupe d'experts de l'AEN, 1981)

£2.80   US$7.00   F28,00

Siting of Radioactive Waste Repositories in Geological Formations
(Proceedings of the Paris Workshop, 1981)

Choix des sites des dépôts de déchets radioactifs dans les formations géologiques
(Compte rendu d'une réunion de travail de Paris, 1981)

£6.80   US$15.00   F68,00

Near-Field Phenomena in Geologic Repositories for Radioactive Waste
(Proceedings of the Seattle Workshop, United States, 1981)

Phénomènes en champ proche des dépôts de déchets radioactifs en formations géologiques
(Compte rendu de la réunion de travail de Seattle, Etats-Unis, 1981)

£11.00 $24.50   F110,00

Disposal of Radioactive Waste — An Overview of the Principles Involved, 1982

Évacuation des déchets radioactifs — un aperçu des principes en vigueur, 1982

Free on request — Gratuit sur demande

Geological Disposal of Radioactive Waste — Research in the OECD Area (1982)

Évacuation des déchets radioactifs dans les formations géologiques — Recherches effectuées dans les pays de l'OCDE (1982).

Free on request — Gratuit sur demande

• • •

SAFETY

SÛRETÉ

Safety of Nuclear Ships
(Proceedings of the Hamburg Symposium, 1977)

Sûreté des navires nucléaires
(Compte rendu du Symposium de Hambourg, 1977)

£17.00   US$35.00   F140,00

Nuclear Aerosols in Reactor Safety
(A State-of-the-Art Report by a Group of Experts, 1979)

Les aérosols nucléaires dans la sûreté des réacteurs
(Rapport sur l'état des connaissances établi par un Groupe d'Experts, 1979)

£8.30   US$18.75   F75,00

Plate Inspection Programme
(Report from the Plate Inspection Steering Committee — PISC — on the Ultrasonic Examination of Three Test Plates), 1980

Programme d'inspection des tôles
(Rapport du Comité de Direction sur l'inspection des tôles — PISC — sur l'examen par ultrasons de trois tôles d'essai au moyen de la procédure «PISC» basée sur le code ASME XI), 1980

£3.30   US$7.50   F30.00

Reference Seismic Ground Motions in Nuclear Safety Assessments
(A State-of-the-Art Report by a Group of Experts, 1980)

Les mouvements sismiques de référence du sol dans l'évaluation de la sûreté des installations nucléaires
(Rapport sur l'état des connaissances établi par un Groupe d'experts, 1980)

£7.00   US$16.00   F64,00

Nuclear Safety Research in the OECD Area. The Response to the Three Mile Island Accident (1980)

Les recherches en matière de sûreté nucléaire dans les pays de l'OCDE. L'adaptation des programmes à la suite de l'accident de Three Mile Island (1980)

£3.20   US$8.00   F32,00

Safety Aspects of Fuel Behaviour in Off-Normal and Accident Conditions
(Proceedings of the Specialist Meeting, Espoo, Finland, 1980)

Considérations de sûreté relatives au comportement du combustible dans des conditions anormales et accidentelles
(Compte rendu de la réunion de spécialistes, Espoo, Finlande, 1980)

£12.60 $28.00   F126,00

Safety of the Nuclear Fuel Cycle (A State-of-the-Art Report by a Group of Experts, 1981)

Sûreté du Cycle du Combustible Nucléaire
(Rapport sur l'état des connaissances établi par un Groupe d'Experts, 1981)

£6.60 $16.50   F66,00

Critical Flow Modelling in Nuclear Safety
(A State-of-the-Art Report by a Group of Experts, 1982) [in preparation]

La modélisation du débit critique et la sûreté nucléaire
(Rapport sur l'état des connaissances établi par un Groupe d'Experts, 1982) [en préparation].

£6.60 $13.00   F66,00   DM 33.00

•   •   •

LEGAL PUBLICATIONS — PUBLICATIONS JURIDIQUES

Convention on Third Party Liability in the Field of Nuclear Energy — incorporating the provisions of Additional Protocol of January 1964

Convention sur la responsabilité civile dans le domaine de l'énergie nucléaire — Texte incluant les dispositions du Protocole additionnel de janvier 1964

Free on request — Gratuit sur demande

Nuclear Legislation, Analytical Study: "Nuclear Third Party Liability" (revised version, 1976)

Législations nucléaires, étude analytique: "Responsabilité civile nucléaire" (version révisée, 1976)

£6.00   US$12.50   F50,00

Nuclear Legislation, Analytical Study: "Regulations governing the Transport of Radioactive Materials" (1980)

Législations nucléaires, étude analytique: "Réglementation relative au transport des matières radioactives" (1980)

£8.40   US$21.00   F84,00

Nuclear Law Bulletin
(Annual Subscription — two issues and supplements)

Bulletin de Droit Nucléaire
(Abonnement annuel — deux numéros et suppléments)

£6.00 $13.00   F60,00

Index of the first twenty five issues of the Nuclear Law Bulletin

Index des vingt-cinq premiers numéros du Bulletin de Droit Nucléaire

Description of Licensing Systems and Inspection of Nuclear Installation (1980)

Description du régime d'autorisation et d'inspection des installations nucléaires (1980)

£7.60   US$19.00   F76,00

NEA Statute

Statuts de l'AEN

Free on request — Gratuit sur demande

• • •

OECD SALES AGENTS
DÉPOSITAIRES DES PUBLICATIONS DE L'OCDE

ARGENTINA – ARGENTINE
Carlos Hirsch S.R.L., Florida 165, 4° Piso (Galería Guemes)
1333 BUENOS AIRES, Tel. 33.1787.2391 y 30.7122

AUSTRALIA – AUSTRALIE
Australia and New Zealand Book Company Pty, Ltd.,
10 Aquatic Drive, Frenchs Forest, N.S.W. 2086
P.O. Box 459, BROOKVALE, N.S.W. 2100

AUSTRIA – AUTRICHE
OECD Publications and Information Center
4 Simrockstrasse 5300 BONN. Tel. (0228) 21.60.45
Local Agent/Agent local :
Gerold and Co., Graben 31, WIEN 1. Tel. 52.22.35

BELGIUM – BELGIQUE
LCLS
35, avenue de Stalingrad, 1000 BRUXELLES. Tel. 02.512.89.74

BRAZIL – BRÉSIL
Mestre Jou S.A., Rua Guaipa 518,
Caixa Postal 24090, 05089 SAO PAULO 10. Tel. 261.1920
Rua Senador Dantas 19 s/205-6, RIO DE JANEIRO GB.
Tel. 232.07.32

CANADA
Renouf Publishing Company Limited,
2182 St. Catherine Street West,
MONTRÉAL, Que. H3H 1M7. Tel. (514)937.3519
OTTAWA, Ont. K1P 5A6, 61 Sparks Street

DENMARK – DANEMARK
Munksgaard Export and Subscription Service
35, Nørre Søgade
DK 1370 KØBENHAVN K. Tel. +45.1.12.85.70

FINLAND – FINLANDE
Akateeminen Kirjakauppa
Keskuskatu 1, 00100 HELSINKI 10. Tel. 65.11.22

FRANCE
Bureau des Publications de l'OCDE,
2 rue André-Pascal, 75775 PARIS CEDEX 16. Tel. (1) 524.81.67
Principal correspondant :
13602 AIX-EN-PROVENCE : Librairie de l'Université.
Tel. 26.18.08

GERMANY – ALLEMAGNE
OECD Publications and Information Center
4 Simrockstrasse 5300 BONN Tel. (0228) 21.60.45

GREECE – GRÈCE
Librairie Kauffmann, 28 rue du Stade,
ATHÈNES 132. Tel. 322.21.60

HONG-KONG
Government Information Services,
Publications/Sales Section, Baskerville House,
2/F., 22 Ice House Street

ICELAND – ISLANDE
Snaebjörn Jönsson and Co., h.f.,
Hafnarstraeti 4 and 9, P.O.B. 1131, REYKJAVIK.
Tel. 13133/14281/11936

INDIA – INDE
Oxford Book and Stationery Co. :
NEW DELHI-1, Scindia House. Tel. 45896
CALCUTTA 700016, 17 Park Street. Tel. 240832

INDONESIA – INDONÉSIE
PDIN-LIPI, P.O. Box 3065/JKT., JAKARTA, Tel. 583467

IRELAND – IRLANDE
TDC Publishers – Library Suppliers
12 North Frederick Street, DUBLIN 1 Tel. 744835-749677

ITALY – ITALIE
Libreria Commissionaria Sansoni :
Via Lamarmora 45, 50121 FIRENZE. Tel. 579751/584468
Via Bartolini 29, 20155 MILANO. Tel. 365083
Sub-depositari :
Ugo Tassi
Via A. Farnese 28, 00192 ROMA. Tel. 310590
Editrice e Libreria Herder,
Piazza Montecitorio 120, 00186 ROMA. Tel. 6794628
Costantino Ercolano, Via Generale Orsini 46, 80132 NAPOLI. Tel.
405210
Libreria Hoepli, Via Hoepli 5, 20121 MILANO. Tel. 865446
Libreria Scientifica, Dott. Lucio de Biasio "Aeiou"
Via Meravigli 16, 20123 MILANO Tel. 807679
Libreria Zanichelli
Piazza Galvani 1/A, 40124 Bologna Tel. 237389
Libreria Lattes, Via Garibaldi 3, 10122 TORINO. Tel. 519274
La diffusione delle edizioni OCSE è inoltre assicurata dalle migliori
librerie nelle città più importanti.

JAPAN – JAPON
OECD Publications and Information Center,
Landic Akasaka Bldg., 2-3-4 Akasaka,
Minato-ku, TOKYO 107 Tel. 586.2016

KOREA – CORÉE
Pan Korea Book Corporation,
P.O. Box n° 101 Kwangwhamun, SÉOUL. Tel. 72.7369

LEBANON – LIBAN
Documenta Scientifica/Redico,
Edison Building, Bliss Street, P.O. Box 5641, BEIRUT.
Tel. 354429 – 344425

MALAYSIA – MALAISIE
and/et SINGAPORE - SINGAPOUR
University of Malaya Co-operative Bookshop Ltd.
P.O. Box 1127, Jalan Pantai Baru
KUALA LUMPUR. Tel. 51425, 54058, 54361

THE NETHERLANDS – PAYS-BAS
Staatsuitgeverij
Verzendboekhandel Chr. Plantijnstraat 1
Postbus 20014
2500 EA S-GRAVENHAGE. Tel. nr. 070.789911
Voor bestellingen: Tel. 070.789208

NEW ZEALAND – NOUVELLE-ZÉLANDE
Publications Section,
Government Printing Office Bookshops:
AUCKLAND: Retail Bookshop: 25 Rutland Street,
Mail Orders: 85 Beach Road, Private Bag C.P.O.
HAMILTON: Retail Ward Street,
Mail Orders, P.O. Box 857
WELLINGTON: Retail: Mulgrave Street (Head Office),
Cubacade World Trade Centre
Mail Orders: Private Bag
CHRISTCHURCH: Retail: 159 Hereford Street,
Mail Orders: Private Bag
DUNEDIN: Retail: Princes Street
Mail Order: P.O. Box 1104

NORWAY – NORVÈGE
J.G. TANUM A/S Karl Johansgate 43
P.O. Box 1177 Sentrum OSLO 1. Tel. (02) 80.12.60

PAKISTAN
Mirza Book Agency, 65 Shahrah Quaid-E-Azam, LAHORE 3.
Tel. 66839

PHILIPPINES
National Book Store, Inc.
Library Services Division, P.O. Box 1934, MANILA.
Tel. Nos. 49.43.06 to 09, 40.53.45, 49.45.12

PORTUGAL
Livraria Portugal, Rua do Carmo 70-74,
1117 LISBOA CODEX. Tel. 360582/3

SPAIN – ESPAGNE
Mundi-Prensa Libros, S.A.
Castelló 37, Apartado 1223, MADRID-1. Tel. 275.46.55
Libreria Bosch, Ronda Universidad 11, BARCELONA 7.
Tel. 317.53.08, 317.53.58

SWEDEN – SUÈDE
AB CE Fritzes Kungl Hovbokhandel,
Box 16 356, S 103 27 STH, Regeringsgatan 12,
DS STOCKHOLM. Tel. 08/23.89.00

SWITZERLAND – SUISSE
OECD Publications and Information Center
4 Simrockstrasse 5300 BONN. Tel. (0228) 21.60.45
Local Agents/Agents locaux
Librairie Payot, 6 rue Grenus, 1211 GENÈVE 11. Tel. 022.31.89.50

TAIWAN – FORMOSE
Good Faith Worldwide Int'l Co., Ltd.
9th floor, No. 118, Sec. 2
Chung Hsiao E. Road
TAIPEI. Tel. 391.7396/391.7397

THAILAND – THAILANDE
Suksit Siam Co., Ltd., 1715 Rama IV Rd,
Samyan, BANGKOK 5. Tel. 2511630

TURKEY – TURQUIE
Kültur Yayinlari Is-Türk Ltd. Sti.
Atatürk Bulvari No : 77/B
KIZILAY/ANKARA. Tel. 17 02 66
Dolmabahce Cad. No : 29
BESIKTAS/ISTANBUL. Tel. 60 71 88

UNITED KINGDOM – ROYAUME-UNI
H.M. Stationery Office, P.O.B. 569,
LONDON SE1 9NH. Tel. 01.928.6977, Ext. 410 or
49 High Holborn, LONDON WC1V 6 HB (personal callers)
Branches at: EDINBURGH, BIRMINGHAM, BRISTOL,
MANCHESTER, BELFAST.

UNITED STATES OF AMERICA – ÉTATS-UNIS
OECD Publications and Information Center, Suite 1207,
1750 Pennsylvania Ave., N.W. WASHINGTON, D.C.20006 – 4582
Tel. (202) 724.1857

VENEZUELA
Libreria del Este, Avda. F. Miranda 52, Edificio Galipan,
CARACAS 106. Tel. 32.23.01/33.26.04/33.24.73

YUGOSLAVIA – YOUGOSLAVIE
Jugoslovenska Knjiga, Terazije 27, P.O.B. 36, BEOGRAD.
Tel. 621.992

Les commandes provenant de pays où l'OCDE n'a pas encore désigné de dépositaire peuvent être adressées à :
OCDE, Bureau des Publications, 2, rue André-Pascal, 75775 PARIS CEDEX 16.

Orders and inquiries from countries where sales agents have not yet been appointed may be sent to:
OECD, Publications Office, 2 rue André-Pascal, 75775 PARIS CEDEX 16.

65579-9-1982

PUBLICATIONS DE L'OCDE, 2, rue André-Pascal, 75775 PARIS CEDEX 16 - N° 42357 1982
IMPRIMÉ EN FRANCE
(66 82 06 3) ISBN 92-64-02351-8